都市校园

Urban Learning Typologies

[丹] BIG建筑事务所等 | 编

司炳月 张晗 王晓华 | 译

大连理工大学出版社

都市校园

都市校园

办公场所

交织于历史叙事中

Urban Learning Typologies

To Work

Weaving Together Narratives of History

信息、人和不断变化的优先事项
Information, people and changing a few priorities

Philip D. Plowright

如果明知道一些事情重要，但却选择置之不理会怎样？究竟是由于道德低下、能力有限，还是傲慢自大？如果是建筑设计的物理问题呢？例如，不考虑气候、洪涝区识别、太阳的朝向问题，或者指定使用不合格的材料，结果导致财产损失或者产生意想不到的建筑维修费用。几乎可以肯定地说，这纯属道德问题，也可能是法律问题。但是，如果是其他类型信息的问题呢？我这样问，既不是纸上谈兵，也不是信口开河。

信息很重要，因为建筑设计的核心就是要将数据转化为可操作的信息，从而决定建筑的形式构成。从数字工具中使用参数的这一趋势、算法对空间规划的影响以及对实证设计的需求中可以看出，信息的重要性越来越明显。但是其中不可告人的秘密是：这些过程并不像我们认为的那样新。它们使用与"传统设计"相同的信息，只不过工作方式已从模拟化更改为数字化——将选定的类别放在最前面，更清晰地介绍人们的需求，并准确识别何时何地使用了哪些信息。

虽然这些趋势仍处于襁褓中，但未来一片光明，不过我这么说的同时也要提出几点说明。积极的一面是决策的过程是可见的，而且不同类型信息可能产生怎样不同的形式效果，也是可见的。无论是基于事实还是基于合理的信念。有担心只是因为我们通过结构化数据来做决策并不能保证数据总是相关的，也不能保证我们会知道如何运用这些数据。我们还需要克服一些建筑学中的历史偏见。

What if we knew something important and then choose to ignore it? Would this be an ethical issue, or an issue of competence, or an issue of hubris? If I was talking about a physical aspect of building design – ignoring climate issues, flood zone identification, or solar orientation, for example, or specifying inadequate materials which later caused loss of property or incurred unexpected building maintenance costs – it most certainly would be ethical and probably also legal. But what about other types of information? I am not asking this question theoretically nor am I asking it casually.

Information matters because translating data into actionable information to shape formal composition is the core of architectural design. This is made more visible in the current trends of parametrics as used in digital tools, the influence of algorithms in space planning and the call for evidence-based design. The dirty secret is that these processes are not as new as we think. They use the same information as "traditional design" but simply move from analogue to digital workflows – frontloading the selection of categories, introducing the need for more clarity and requiring an explicit identification of what, where and when information is being used.

I see these trends as overwhelmingly positive although they are still in their infancy, although I say this with a few caveats. The positive is the visibility created for making decisions as well as how different types of information might create different formal effects – whether based on things that are factual or about justified belief. The concern is just because we structure data to make decisions doesn't mean that the data are always relevant or that we know how

问题是，我们优先考虑的是物体和形状，而不是用来证明这些物体和形状的信息，这根本就是把结果和目的混为一谈。作为设计师，我们通过形象而非经验去理解物体，因为建筑不是与原型一一对应的，而是通过设计师提出的模型来体现的。这意味着我们倾向于优先考虑物体的形象，而不考虑其社会背景。我们在企图将"参数"变成"参数主义"的过程中就已经看到了这一点，实际上就是试图使用蛮力将工具应用转化为视觉应用。

我可以再说几个其他问题，例如：只是在象征意义上将建筑诠释成雕塑作品；使用形象来创造以新颖为基础的社会地位；或者将建筑定位为纯艺术（而不是关键的设计），这意味着，最终，建筑要么是表现自我，要么是批判社会，要么是负责表达进步。然而，这些问题本身就会自行展开一场完整的讨论，让我严重跑题。

在建筑中，有一类关键信息总是莫名缺失，那就是人和他们占据空间的方式。当然，我们会倾听客户的喜好，我们甚至可能会使用一些基本的研究工具，例如，进行访谈，但是这些都是在具有相当大规模的需求情况下进行的，或者说是建立共情的一种方式。那么人们了解环境的方式是如何的呢？在特定地点，人类认知进行交互的方式又是如何产生统一的含义的呢？

目前，这一内容没有在建筑学中以任何形式化的方式存在，即使有，也是在实践中发展起来的，

to apply them. There are some historical biases that exist within architecture that need to be overcome.

Our problem is that we prioritize objects and shapes over the information we used to justify those objects and shapes – essentially mistaking the outcome for the purpose. As designers, we then understand those objects through their image rather than through the experience since architecture always works through proxies of what we propose rather than one-to-one prototypes. This means that we tend to prioritize the image of the object independent to its social context – we have seen this in the attempt to turn parametics into "parametricism", effectively attempting a brute force translation from a tool application to a visual style.

I could go on into other issues – such as the translation of buildings into sculptural objects which are only engaged on a symbolic level; the use of images to create social status based on novelty; or the positioning of architecture as fine art (rather than as critical design) means that, ultimately, architecture is either about self-expression, social critique or responsible for expressions of progress. However, these would spiral into an entire discussion by themselves and get me further from the point I wish to make.

A type of critical information is strangely absent from architecture: people and how we occupy space. Sure, we listen to clients when they tell us what they like and don't like and, perhaps, we might even use some basic research tools like conducting interviews but these are usually based on fairly large-scale needs or desires, or as a way to build empathy. What about how people make sense of environments and how the interaction of human cognition

以有利于抢占市场的专有信息被保存下来。它在学术界是不存在的，也不属于教授建筑原理的基础课。相反，我们教的平面设计或雕塑表达，才是建筑学的基础。

为什么要忽视那些在这个负有为人们创造丰富环境使命的领域里工作的人呢？

原因之一可能是一种无意识的偏见，这种偏见会影响到我们对可以操作的信息类型的判断，因为我们认为人是特殊的个体，我们不能也不该预判人的行为。我们知道，每个人喜欢的东西都不一样，或者喜欢一种风格的衣服甚过另一种。我们也理解每种文化都有不同的价值观，等等。我们相信自由意志高于命运，并反对任何与宿命论有关的东西。每个人都有各自的宇宙，对吧？

幸运的是，事实并非如此，否则，语言、文化、社会、设计、通信、艺术、文学、科学等领域是绝对不可能实现的。整个人类体验都是建立在共同的意义基础上的，这些意义源于相同的来源，即我们身体在空间中的相互交流。这类知识被称为"具身"，意思是即使我们不用有意识地思考也能了解事物，并且我们不断地使用这些知识来理解我们的环境。这种知识还是人和周边环境以及与环境（如，建筑）中他人相互交流的基础。

要认真对待这些信息，我们需要改变几个优先顺序。在过去的30年里，认知科学已经建立了一个关于人的知识的大数据库。这些内容大多是潜在的，我们都没有意识到。这种知识几乎没有被空间化、被翻译、被编码，从而为建筑所用（笔者最近出版了《通过人性化打造建筑》这本书，试图从最基

in a particular place produces consistent meaning?

This content does not currently exist in architecture in any formalized way. Or if it does, it has been developed through practice and kept as propriety information, hoarded for a market advantage. It is not present in academia and not used in our foundation courses to teach architectural principles. Instead, we teach either graphic design or sculptural expression as the basis of architecture.

Why do we ignore people in a discipline whose responsibility is to make enriched environments for... people?

One reason could be an unconsidered bias that affects what type of information we believe is actionable due to a belief that people are special and individuals and we cannot, or should not, predict human behaviour. We point to the fact that people like different things or prefer one type of clothing over another. We understand that different cultures have different values and so on. We believe in free will over fate and push back against anything that feels deterministic. Each person is a cosmos onto themselves, right?

Well, no. Luckily, this simply is not the case – or language, culture, society, design, communication, art, literature, science (and so on) would simply not be possible. The entirety of human experience is built on a shared basis of meaning that starts from the same source – the interaction of our body in space. This type of knowledge is called embodied, which means that we know things without conscious thought, and we use this knowledge constantly to make sense of our context. This knowledge is also the basis of how people interact with our environment and with each other in that environment (i.e. architecture).

In order to start to take this information seriously, we need to change a few priorities.

In the last thirty years, the cognitive sciences have developed a large database of knowledge about people. Most of this content is latent, things that operate below our attention level. Almost none of this information has been spatialized, translated and coded for use in

本的层面上做到这一点）。这些信息都没有作为一种共享的语言系统地进入学术界或实践中。

建筑学必须承认，大多数的人类信息既不是确定的，也不是完全随机的。这是一种错误的二分法。反之，人类很大程度上会在特定情况下做出相应的反应，虽然不是总是，也不绝对是这样，但大部分是这样。该信息仅仅是可能性而不是确定性（这也是一件好事）。但是，由于每次发生的事情，我们都无法以同样的方式准确预测，因此我们的科学传统就判定此知识无效。我们了解人类的欲望、需求和价值观，只要观察营销和零售设计就可以了解到，空间中的人是可预测的，但建筑不会放弃对个性的狂热追求。我们是不是应该思考一下这个问题？

我们不应一味地关注意识形态的大规模文化价值，而应注意其根本。另外，由于人类用来认知世界的具身思维结构是我们构建许多其他类型意义的重要基础，所以这些根本要素应该要以人为本。

未来10年，我希望，通过人类和我们生存的世界已经建立起来的关系中的共享算法，引入人类对环境的解释，从而改变我们基本的优先顺序。只有这样，建筑设计才会在经验意图的基础上变得更加巩固，不会出现学生要费力取悦指导者的情况，而指导者的意见可能只是出于个人的审美取向或一时的兴起。同时，应该培养出更多有更多的机会向客户和公众传播知识的建筑从业者，让建筑在基础设施和文化发展这两个领域中起到重大作用。

architecture (my latest book, *Making Architecture Through Being Human*, attempts to do this at a very basic level). None of this information has found its way into academia or practice as a shared language in a systematic way.

Architecture needs to accept that most human information does not exist as either deterministic or completely random. This is a false dichotomy. Rather, humans have strong probabilities of acting in a certain way given a certain situation. Not always and not absolutely but mostly. This information addresses potentiality rather than determinism (this is also a good thing). However, because we cannot predict that something happens every time, exactly the same way – our scientific legacy says that knowledge is not valid. We know lots of things about people, their desires, needs and values – we only need to look at marketing and retail design to understand that people in spaces are predictable but architecture maintains its cult of the individual. Maybe we should think about this?

We need to pay more attention to fundamentals rather than always jumping to the large-scale cultural values of ideology. These fundamentals should be based on people through embodied thinking structures that humans use to make sense of the world as they form an important foundation on which we build many, many other types of meaning.

I am hoping the next decade shows a shift in our foundational priorities introducing human interpretation of environments through the shared algorithms that have developed biologically in a relationship between humans and the world we live in. This will allow architectural design to become more defensible based on intentions of experience rather than a situation where students struggle to satisfy the opinion of their instructor, which might only be based on personal aesthetic or whim. It should also produce practitioners who have more access to describing their knowledge to clients and the public, reinforcing architecture as an important and critical player in both infrastructure and cultural development.

都市校园

Urban Learni

能够营造一种成功的学习环境靠什么？周围环境？独特的教室设计？美学表现形式？学校是我们传统建筑环境的基本要素，它不仅服务于学生和教师，也服务于广大社会群体。[1]

很明显，在当代的生活中，我们所造、所用、所行的一切都需要好的设计，这在学校的设计上体现得最为明显。英国等国家的许多研究都表明，学校的设计与学生的表现和行为有明显的联系。虽然仅仅靠好的设计并不能提高学生的水平，但是设计感差的学校会影响教学质量、学生的抱负和自信心以及学校的可持续发展。[2]

那么校园建筑如何提升教学质量，又如何优化学生体验呢？从建筑学的角度来说，学校应该不仅仅是

What makes a successful learning environment? The context? Particular classroom design? Aesthetic expression? Schools are fundamental elements of our conventional architectural environment, catering not only for students and teachers but also for the wider community.[1]

In contemporary life, it is evident that good design is crucial to all that we make, use and do. This could not be more obvious than in school design. Numerous research papers, both in the UK and abroad, demonstrate that there is a clear association between well-designed schools and student performance and behavior. Good design by itself does not elevate standards, yet poorly designed schools impact on the quality of teaching, the ambition and self-confidence of pupils, and the sustainability of a school.[2]

How can school architecture improve education and the student experience? Architecturally, a school should be more than the sum of its parts. How do spaces change to accommodate a variety

ng Typologies

各个空间的结合。随着时间的推移，空间要发生怎样的变化，才能适应拥有不同层次的细节和批判性思维的教育需求？学校应成为特殊的学习场所，促进儿童进步、身心健康、学业提高，并鼓励教师和学生积极地互动。学校不仅是白天上课的地方，也是课后儿童保育空间、体育俱乐部、娱乐活动场所和当地社区功能空间。简而言之，新学校的设计需要以灵活性和适应性为核心。

大多数新学校的选址都在拥挤的城市里，所以设计难度很大。建筑师如何才能设计建造出既满足当前需求，又满足未来需求的开放、透明、灵活、适应性强的空间呢？新的教学方法正在改变学校设计的建筑组成，小组活动室、工作室、便于独立学习或实践活动的区域以学生为主来设计，而非以教师为主。[3]

of educational demands over time, with different levels of detail and critical thinking? Schools should be exceptional places to attend and study in, nurturing children's advancement, heath, psychology and learning, as well as encouraging positive interactions between teachers and fellow students. They are places not only for daily lessons, but for after-school childcare, sports clubs, recreational activities, and local community functions. In short, new schools need to be designed with flexibility and adaptability at the core.

In most cases, the sites for new schools are difficult and in dense urban areas. How do architects design and construct open, transparent, resilient and adoptable spaces that are responsive both to the needs of the present and the future? New pedagogical approaches are altering the architectural composition of the school design; small group settings, studios and areas for independent studying and practical activities place the priority on a student-focused approach rather than a teacher-focused one.[3]

都市校园
Urban Learning Typologies

Gihan Karunaratne

人口快速增长、社会老龄化、移民增加以及激增的经济和社会基础设施需求等问题，是发达国家和发展中国家许多地方当局正面临的挑战。其中，城市委员会和规划当局面临的问题是缺乏建造新学校的资金、规划和设计。土地资源不足和城市密集化，必然导致许多学校的设计和建造条件严重受限。

通常，主要城市土地用于住宅开发和商业开发。学校的新校址通常位于棕地，与高速公路或铁路基础设施相邻，且与生源数量大于学校容量的学校聚集区相距甚远。

这些新学校要解决的问题是想办法将外部空间容纳进来，以使休息区域、体育训练区域、室外学习区域和社交空间等有意义地联系起来。在垂直学校类型中，应通过更具创新性的设计方法，减少对安全问题和冷漠忽视的担忧。为了避免拥挤和学生之间的肢体冲突，经深思熟虑而设计的垂直交通路线和楼梯必不可少。

如今，建筑师们正在设计一种带有混合项目的垂直学校，这些项目不仅提供学习设施，还可以容纳许多当地的社区活动。在英国，垂直学校的设计要保证在假期可供社区和其他机构使用，这样一来就给学校提供了基本的收入来源。

这些建筑的主要元素是主楼梯、露台和垂直连接。通常，起连接作用的公共空间和垂直露台可用以开展重要的室内或室外会议和教学，以鼓励协作和学习。[4]

问题是，如何更好地通过有效、经济、可持续方式来设计学校环境，并同时在市中心创造富有启发性的教育空间?

近期研究表明，创新型的校园建筑和教室设计可以优化教学效果和学生的学习体验。但是，在设计学校时需要考虑未来的学习环境和教育项目。建筑将如何预见未来学习的需求呢?

从广义上讲，学校建筑的空间设计就是避免“所有空间都一个尺寸”的类型。这种传统的学校建筑类型多年来备受争议，特别是20世纪50年代和70年代的开放式布局，在这个讨论中不断被提及。[5]

A rapid increase in population growth, an ageing society, a rise in immigration and an upsurge in demands for economic and social infrastructure are challenges that many urban local authorities face in developed as well as developing nations. Among the issues confronting urban councils and planning authorities is the funding, planning and design of new schools. The scarcity of sufficient land and urban densification inevitably means that many schools are designed and built in very constricted conditions.

Typically, prime urban land is allocated to housing and commercial developments. New sites for schools are commonly on brownfield sites, often adjacent to motorways or railway infrastructure and frequently remote from the school catchment areas where demand exceeds the supply of school places.

The challenge for these new schools is finding ways to incorporate external spaces so that areas for break time, physical education, external learning and social spaces are meaningfully connected. In the vertical school typology these challenges require a particularly innovative design approach to respond to concerns over safety and unwelcome overlooking. Carefully considered vertical circulation routes and well-designed staircases are essential to avoid congestion and conflict.

Today, architects are designing vertical schools with hybrid programs which not only provide learning facilities but also accommodate a number of local community activities. In the UK vertical schools are designed to be used beyond the school working day by community and other organisations, access which provides an essential revenue stream.

Primary staircases, terracing and vertical connectivity are the main elements of such buildings. Often the connecting public spaces and vertical terraces offer vital indoor or outdoor meeting and teaching which encourage collaborative and didactic learning.[4]

The question that arises is how best to design school environments in an effective, economic and sustainable way while creating an inspirational educational space in the inner city.

Recent studies have shown that innovative school architecture and classroom design can improve teaching effectiveness and the student learning experience. However, schools need to be designed with consideration for future of learning environments and educational programs. How will architecture anticipate the needs of future learning?

In broad terms, the tendency in the spatial design of school buildings is to move away from the "one size for all" prototype. This traditional school building typology has been debated for many years, a particular focus being

20世纪80年代，奥斯卡·尼迈耶在巴西里约热内卢为公共教育综合中心（CIEP）建立了一份标准化蓝图。尼迈耶的实验教育项目包括一套基于混凝土构件的系统，该系统还包括建造一间综合的预制工厂。[6]

尼迈耶项目体系中的学校是以一组组由单独的建筑组成的建筑群的形式布置的。正如里约热内卢世界遗产研究所所长华盛顿·法贾尔多所说："有了基础建设，然后就会有一系列相关的建筑：一座室外有顶体育馆、一座八角形的图书馆和为住宿的小学生提供的屋顶宿舍。"[7]

该方案对建筑的空间构成、规模和连接方式进行了复杂的探讨，而且对原有的教学模式发起挑战，建筑的每一个细节都经过深思熟虑，创造出的空间让学生觉得他们就是这里的未来住户，而不是受教育和被控制的个体。[8] "尼迈耶方案的另一个成就是，标准化所节省的资金可能会补贴到课程中，所以学校能够提供从上午7点到晚上10点的全日制课表。CIEP项目重视学生文化课的同时也注重学生自身素质的培养。"[6]

随着学校真实环境建设的发展，学校的术语也发生了相当大的变化：现在用"学习空间"来代替教室。如今的许多学校类似于现代工作场所，具有更灵活的时间性和空间性，同时能够兼顾"个性化学习"方法，取代了20世纪传统的"小房间和铃声"式的学校模式。[9] 新的学校内部不再有制度化的氛围，而是更有"家"的感觉。内部和外部空间的连接区域为散步和偶然的邂逅创造出活动空间，为学习和社交创造了公共、私密和亲密的空间。在谷歌、脸书和其他现代办公模式中都可以找到类似的空间类型。

由BIG建筑事务所设计的海茨学校（70页）位于一片与华盛顿特区接壤的密集的城市区域中，在一个非常狭窄的地块上安营扎寨，这个地块被三条公路和罗斯林高地公园包围。该项目给阿灵顿街区的密度带来了变化，学校的建筑群成为市中心的休息区。学校设计有景观和露台，面向并延伸至学校的机构空间，开始是休闲娱乐区的绿色草坪，然后向上攀升到屋顶阁

the open-plan layouts of the 1950s and 1970s.[5]

In the 1980s, in Rio de Janeiro, Brazil, Oscar Niemeyer established a standardised blueprint for The Integrated Center of Public Education (CIEP). Niemeyer's experimental educational project comprised a system based on a bare kit of concrete elements, which included the creation of an integrated factory for prefabrication.[6]

The schools in Niemeyer's system were arranged as clusters of individual buildings. As Washington Fajardo, president of the Rio World Heritage Institute, put it: "You have the basic building, and then you have a series of associated buildings: an outdoor covered sports hall, an octagon-shaped library and a house on the roof for live-in pupils."[7]

The scheme is a complex essay in architectural spatial composition that plays with scale and connections, and challenges pedagogical models, every detail of the architecture has been considered to make spaces where the occupants feel that they are treated as future citizens, rather than as subjects to be educated and controlled.[8]

"One result of the Niemeyer's scheme is that the money conserved by standardisation may be subsidized into the curriculum. Schools could allow to offer a fulltime curriculum obtainable from 7a.m. to 10p.m. CIEP programs not only value pupils' cultures as well as strengthen them."[6]

With the development of the physical educational environment, there has been a comparable change in terminology: in lieu of classrooms there are now "learning spaces", in place of the traditional "cells and bells" model of the 20th century. Many schools now resemble modern workplaces with a certain amount of flexibility of time and space, incorporating a more "personalised learning" approach.[9] In place of an institutional atmosphere, new school interiors have a more "homely" feel. Connections between interior and exterior spaces create networks of movements, for promenading and chance encounters, and create public, private and intimate spaces for learning and socialising. Similar spatial typologies can be found at Google, Facebook and other modern office models.

The Heights School by BIG (p.70) is located on a tight plot in dense urban area bordering Washington, D.C., surrounded by three roads and edges of the city's Rosslyn Highlands Park. The density of the urban Arlington neighborhood is transformed by the project. The school complex is an urban sanctuary. The landscape and terraces open and extend into the institutional spaces of the school, initially as the green lawn on the recreational field which then ascends up to the rooftop gardens. This greenery is a linking factor of the broad design.

楼。绿化部分是整个设计的连接元素。

可以将建筑的语言理解为学校的各个楼层像一个个薄片一样呈扇形展开，中间容纳了一个公共的旋转空间，它们作为建筑构件展现在人们眼前，形成不同的形式和空间布置。

正如建筑师描述的那样，由此而形成的连续的露台式空间通过一部弯曲的楼梯紧密相连，旋转楼梯贯穿建筑的所有楼层，连接起建筑的室内和室外空间，将听不同课程、不同年龄段的所有学生在视觉上、实际上联系起来。使用者可以亲眼看到周围发生的事情，自由地出入多个空间，同时感觉到与周围紧密地融为一体。

被动监测、远距离视野、定格的风景：每一个露台都有优美的风景，不仅是学生们进行社交的场所，还成为非正式的室外学习空间。连接建筑与室外空间的人行道让学生们一直都能感到与户外零距离。每层楼都通过不同的方式实现了内外空间共存，每层的绿色露台成为教室的延伸部分。这种颇具神秘感的设计给师生们创造了一种室内外学习体验和一种社交氛围。

海茨学校使用了几种十分坚固、自然的材料，给人以雕塑作品的印象。外立面使用了白色的釉面砖，使单一的板块顺畅地衔接在一起。然而，建筑的焦点不是建筑本身，而是建筑内部。在内部，大楼是一个功能性的现代学校，有时建筑所展示的重点会发生变化，这让人想到这不是一座单纯的大楼，而是一座有生命的建筑。

玻璃立面呈现通透质感，与外部在视觉上连接起来。空间明亮通风，室内外过渡流畅自然，屋顶花园也是设计中至关重要的一部分。[10]

BIG在设计海茨学校时，力求保留非叙事性的学校建筑特点，同时追求垂直配置。[11] 螺旋式楼梯将各个楼层连接起来，这样教师和学生在学校内走动的同时，在视觉上也与周围环境产生联系。楼层越高，就越安静祥和，上层是集中学习和安静放松的空间，下层则是互动和游戏的区域。楼梯旁边的每一条走廊都有其独特的配色方案，给缺少日照的入口区域带来轻松明快的特点。

The architectural lexicon can be understood by perceiving the school's different levels as slices that fan out encompassing a common rotation point, unfolding as architectural elements into forms and arrangements of spaces.

As described by the architects, the resultant cascading terraces are closely knit by a curving stair that weaves through all levels – inside as well as outside – uniting all students, from both programs and all ages, visually and physically connected to each other. The occupants can sense a greater connection to the surroundings when they observe what is happening around them and can move effortlessly from one space to another.

Passive surveillance, long views, and framed landscape: Each terrace is landscaped to lend itself not just to the social life of the students but also as informal outdoor spaces for learning. Using walkways which link the buildings to the exterior, students are always aware that the outside is there. Individual levels are, in a way, constituted of interior and an exterior space. Green terraces above each floor become an extension of the classrooms. This rather enigmatic quality creates an indoor-outdoor learning experience as well as a social landscape for both students and teachers.

The material palette of The Heights School is fairly robust and elemental, enhancing the sculptural effect. The external elevation uses white-glazed brick that dissolves the singular plates into a discursive articulation. Nevertheless, the focal point is not the building itself, but rather what happens inside it. Internally, the complex is a functioning modern school with occasional movements of architectural theater, which serves as a reminder that this is architecture and not just a building.

A glass elevation presents openness and achieves visual connectivity to the outside. Spaces are light and airy, and the boundary between inside and outside remains fluid, for the rooftop gardens are also an integral component of the design.[10]

At The Heights School BIG have sought to retain the sensitivity of non-narrative school building while pursuing a vertical configuration.[11] A spiral stairway connects the different levels, enabling pupils and teachers to circulate throughout the school while remaining visually connected to their surroundings. The higher the level, the more tranquil and peaceful the character: the upper levels contain spaces for concentrated study and quiet relaxation, while lower areas are dedicated to interactive and playful work. Each of the corridors adjoining the stairs has its own color scheme, giving a buoyant, accessible character to access areas not illuminated by daylight.

The lobby, which serves as a setting for community functions, was conceived as a central meeting point. Seating steps have a mixture of purposes: they provide a place to hang out during breaks, an auditorium for assemblies and a playground for varied activities. The Heights School is distinguished by its planning, not its formal expression. Its creative plan solves the multiple demands of site and function practice in a manner which seems

大厅是发挥社区功能的场所，是一个中央会面地点。座位台阶有多种用途：它们可以是一个课间休息场所，可以是集会礼堂，也可以是各种活动的举办场地。海茨学校的独特之处在于它的空间规划，而不是建筑物的形式表达。其创造性的布局解决了场地和功能实践的各种难题，而且解决方式看起来很轻松。经过景观设计的空间模糊了学校室内外空间的边界，并将绿化引入建筑中，更能吸引当地社区居民。

红岭实验小学由源计划建筑师事务所设计（36页），建在中国深圳的填海地上。该场地原规划修建一所有24个班的小学，后因周边学校缺口巨大而增加至36个班，现建筑面积为原规划的两倍。为了充分利用场地，建筑师将学校的教学楼分为高度不同的东西两个部分。

对城市结构、街道语言和行人流线的参考，很大程度上影响了学校建筑群的设计。混凝土的质感增加了空间的深度，使室内、室外和城市空间有了连续性。学校场地北高南低的现状让每层的三排教室从南往北各有1m的高差，以连接不同楼层上的三排学习单元。该设计针对中国南部亚热带气候中高密度的城市环境，囊括了水平且灵活松散排布的教学单元组织和有机的绿植布置，并加入了动态山谷庭院的独特视角。

位于维也纳Hirschstetten区、由PSLA建筑事务所设计的贝雷斯加斯教育园区（54页）是维也纳市计划和实现的一种公私合作模式的建筑。贝雷斯加斯是一个旨在鼓励、促进创新与创造性学习的校园。该学校是调和设计和教育标准的一个实验性尝试。良好的建筑和空间配置能够提高学生的受教育水平、表现、成绩，并且能规范行为。建筑的设计理念就是：创新的建筑和教室设计不仅能提高教育的效率和学生的学习能力，还能提高更大范围的社区的体验。

学校为14岁以下的学生提供了一个综合性的教育学院，其中包括一个体育馆、乐队排练室、户外游戏和体育设施。[12] 在这里，建筑师以鼓励个体学习的方式设计了一个多功能的学院，它在学校和休闲区之间交织，呼应了当代的社会发展和教学思想。学校可以容纳不同规模的团体活动，其中包括维也纳校园模式的自主管理活动和开放式教学活动。为了使工作和休闲

effortless. The landscaped spaces which blur the boundaries of the inside and outside and bring greenery into the building have succeeded in making the school more inviting to the local community.

Hongling Experimental Primary School by O-office Architects (p.36) is designed and constructed on reclamation land in Shenzhen, China. The site was initially programmed for a 24-class school, but in response to the lack of proximate schools, is increased to 36 classes, with double the total floor area. The school building is separated into two sections, varying in height on the east and west sides to fully utilise the site.

There is a reference to urban structure, and the language of the street and pedestrian flows very much informs the design of the school complex. The materiality of the concrete gives depth to the space and creates a continuity of indoor, outdoor and urban space. The height variation of the north and south of the school accommodates a one-meter inclination to connect the three rows of learning units on respective floors.

The design responses to the high-density urban condition in China's semi-tropical southern climate include a horizontal loose cellular fabric and organic greening arrangements, and the incorporation of particular views within the dynamic valley courtyard.

The Education Campus Berresgasse by PSLA Architekten (p.54) in Vienna's Hirschstetten district was planned and realised by the City of Vienna as a public-private partnership model. Berresgasse is a campus which aims to encourage and foster innovation and creative learning; the school is an experiment in reconciling design and education standards. Good architecture and spatial configuration are means to improve students' educational standards, performance, grades and behavior. The idea is that innovative architecture and classroom design can improve teaching effectiveness and student learning, as well as enhance the experience of the wider community.

The campus provides children up to the age of fourteen with a comprehensive, integrated educational academy which includes a gym, band rehearsal room and outdoor play and sports facilities.[12] Here, the architect has designed a multi-use academy which interweaves school and leisure units, and which responds to contemporary societal developments and pedagogical thinking in a manner that supports individual learning. The school accommodates variously-sized group activities, including self-administered and open schooling based on the Viennese Campus Model. Areas are malleable in order to switch swiftly between work and leisure and thus produce most the favorable infrastructure for the project's learning and leisure phases.[13]

The main elements of the architecture are dedicated to open areas, gathering areas and independent study spaces. The campus is arranged into a progression of conventionally orderly learning spaces, each of which has a separate entrance and is clearly legible in its urban terrain. The complex is arranged into a base-like ground floor accommodating four two-story learning clusters of the kindergarten and primary school, which occupy the first

之间能快速切换，建筑师设计了可调整的学校区域，从而为学习和休闲提供了最有利的基础设施。[13]

建筑的主要元素是开放区、聚集区和独立学习空间。建筑师对校园传统有序的学习空间进行了连续布置，每个学习空间都有一个单独的入口，在城市地形中清晰可辨。建筑群被布置成一个类似于基座的一层，包含四个两层的学习区，幼儿园和小学分别位于二层和三层，两个水平布局的新中学学习空间位于四层。这样的设计产生了共约2000m²的露台，这些宽敞的露台位于上层，相互连接，其中一些露台上还设有绿植丰富的小岛。这种空间类型通过创造从室内空间进入的室外教室，重点关注室外学习区域的使用。

贝雷斯加斯学校可以被看作多种建筑类型相互作用的产物，它是联排建筑、板式结构、桥梁结构和大厅结构的混合体。本案这所学校是第一个建成的如此规模的学校项目，它采用辐射状的风车一样的布置方式，有意识地替代了教室典型的正交形式布局。这样的布局方式能增加空间之间的相互关系，也能改善空间与城市环境的关系。

幼儿园和小学通过位于中央的类似大厅的两层中庭相连，中庭在一天中的不同时间会显出不同的特点。不同"BiBers"区域中庭的中央公共空间都有多种用途，这不但是为了鼓励特定年龄段的孩子之间相互交流，也跨越了不同机构之间的年龄差异，在"BiBer家族"内培养了社区意识和联系感。

这些中庭的高度超过7m，最大限度地扩大了视觉和空间的交互性。同时，在窗帘、座椅长廊、活动隔断和被降至仅2.2m的天花板高度的帮助下，休闲和认真工作之间的联系也得到了增强。在一些公共区域可以欣赏到周围地区的景色，而其他公共区域则有紧邻小学教室的内部阳台。所有教室都是按照对角线布置的，而每个星形学习区都能得到日光采光。

建筑体块的表现形式、"跳跃式"窗户的布置、幼儿园小组活动室里游戏室的布局、教室里的定制家具，这一切都确保拥有起伏横竖落叶松木板条立面的教学楼能给师生提供清新、舒适、私密的空间。在贝雷斯加斯学校，建筑已成为交流的背景。学校创造了一个平静空间，让人们摆脱日常生活的琐碎，放缓生活节奏，将美好定格。这是界线的交响曲，能够激发学生探索（空间）和发现（周围景观）的能力。学生可以随时随地学习。这样的环境可以让学生更高效地集中注意力，形成认同感

and second floor levels, and the two horizontal learning clusters of the new middle school, which are positioned on the third floor level. This planning results in a total of around 2,000m² of ample, interwoven terraces at all upper floors, some of which contain islands of rich greenery. It is a spatial typology which focuses on the use of external areas for learning by creating external classrooms which are accessible from the internal spaces.

Berresgasse School can be read as the product of a range of typologies; it is a hybrid of a terraced, slab, bridge and hall structure. This is the first completed school project of this scale that has consciously replaced the orthogonal organisation of the classrooms with a radial, windmill-like configuration that reinforce the interrelationship of the spaces with each other and with the urban setting.

The kindergarten has been bound with the primary school through a central, hall-like two-story atrium with qualities that differ during the day. The central common spaces in the atria of the individual "BiBers" can be used in different ways; their particular territories are intended to encourage age-specific intercommunication but also transcend the wide institutions' age disparities by fostering a sense of community and connectedness within the "BiBer family".

With a height of over seven meters these atria amplify visual and spatial reciprocity while also achieving ample connections for refuge and concentrated work with the assistance of curtains, seating galleries, mobile partitions and "reduced" room heights of just 2.2 meters. Some common areas offer views out into the encompassing surrounds while others have internal balconies that directly adjoin each primary school classroom. All classrooms are presented diagonally, while each of the star-shaped learning clusters is illuminated by daylight.

The expression of the mass, the position of the "jumping" windows, the composition of the playhouses in the kindergarten group rooms, and the fitted furniture in the classrooms ensure that the building, with its fluctuating horizontal and vertical larchwood slatted facade, provides refreshing, domestic, intimate spaces for pupils and teachers. At Berresgasse School architecture becomes a backdrop to the conversation. It has produced a place of calm that offers respite from normal daily routine and slows everything down into a series of moments; it is a symphony of thresholds, and a vehicle of discovery (of spaces) and exploration (of the surrounding landscape). Learning happens everywhere and in all places. The environment allows students to focus more effectively and form a sense of identity and belonging.

The Kalvebod Fælled School by Lundgaard & Tranberg Arkitekter (p.16) is a circular architectural composition in Ørestad, a developing precinct of Copenhagen in Denmark. Its spherical design is as functional as it is symptomatic. The primary intent and philosophical doctrine of the school is to promote physical and mental wellbeing through sports, movement and learning.

The design intends to encourage visual connectivity and movement of its occupiers. The exterior upper-level

和归属感。

卡尔韦博德·费勒德学校由伦高·特兰伯格建筑事务所设计（16页），位于丹麦哥本哈根开发区奥雷斯塔德。这座采用圆形设计的建筑不仅外观独特，而且还具有多种功能。学校的主要目的和设计理念是通过体育运动、积极活动和学习来促进学生们的身心健康。

设计鼓励师生们有更多视觉上的联系，鼓励他们在建筑内多走动。位于建筑物上层的室外教育空间延伸到凸出的学习露台上，这些学习露台通过阳光充足的通高中庭空间联系在一起。学校建筑群的中央还有一座大型体育馆。

建筑的材料和功能得到了一体化设计。设计最大限度地采用自然光、被动通风系统、夜间冷却、有效的生物质供热，以及尽可能使用简单的建筑材料和少量涂料，从而在实现资源可持续的同时，还可以欣赏周围的自然环境，这两点也是设计的重点。更值得注意的是，该建筑的立面被设计成主动表面，可调节角度的穿孔板由使用者手动控制，以避免过多的太阳热量和眩光。

学校为师生提供了社交和教育空间，来促成动态化学习，其中包括咖啡馆和一间重视健康食品与营养的大厨房，这些空间既具有多功能性，又有空间性，符合现今流行的教学理念。

建筑提供了特别的教学设施。放学后，学校又变身为公共空间供当地社区使用。多层空间开放而包容，突破了传统教育体现出的控制、约束，还给学生自由和独立的感觉。该学校设计创造了一种激励人的学习环境，也成为未来城市学校类型建筑的参考样板。

以上四个城市学校建筑案例说明，城市快速发展地区的社区将面临挑战。这些学校项目显示了建筑如何随着时间的推移进行调整，以适应建筑目的和人口结构的变化，进而提高建筑的可持续性。其中三所学校实现了灵活设计，这种灵活性非常必要，不仅能发挥教育所需的各种功能，又向社区开放，让社区居民在任何时候都能在这里集会和进行社交互动，从而促进所有人终身学习。

educational spaces extend into projected study terraces, which are united by a full-height atrium space filled with natural light. At its core, a sizable gymnasium is integrated into the center of the school complex.
The materiality and program of the building are designed in unison. They focus on sustainability and visual aspects of the surrounding natural environment. This is achieved by maximising the use of natural light, passive ventilation, night cooling, efficient biomass heating, and a simple palette of materials to the use of paint where possible. More importantly, the facade is designed as an active surface where shifting perforated panels are controlled by the occupants to avoid heat gain and glare.
Social and educational spaces are provided for teachers and students for dynamic learning. These include cafes and a large kitchen, which underscores the importance of healthy food and nutrition, in the course of multi-functionality and spatial volumes that are pedagogically fashioned.
The building offers exceptional educational facilities. It also serves as a social hub for the local community outside school hours. The layered spaces are open and inviting, offering students a sense of freedom and independence, without the usual control and restrictions of conventional institution models. The school design creates an inspiring learning environment. It also sets an architectural precedent for future urban typologies.
These four examples of the architecture of urban school buildings illustrate the challenges communities facing in areas of rapid urbanisation. They offer solutions which show how architecture can adapt over time, responding to changes in purpose and demographics in order to promote sustainability. The three schools provide the flexibility necessary to ensuring that space can accommodate different functions during the school year and yet be open to the community to host encounters and social interactions that promote learning at all times, by all.

1. Sirkkaliisa Jetsonen, "Setting the Scene for Learning", *The Best School in the World: Seven Finnish Examples from the 21st Century* (Helsinki: Museum of Finnish Architecture, Jun. 2011), pp.72-73
2. Paul Finch, Foreword, *Creating excellent primary schools - A guide for clients* (London: CABE, 2010), p.5
3. Eriika Johansson, Introduction, *The Best School in the World: Seven Finnish Examples from the 21st Century* (Helsinki: Museum of Finnish Architecture, Jun. 2011), pp.8-9
4. Ann Lau, "World Architecture Festival 2016: South Melbourne Primary School", *Dezeen*, Dec. 22, 2016
5. Yael Duthilleul, Alastair Blyth, Wesley Imms and Kristina Maslauskaite, *School Design and Learning Environments in the City of Espoo, Finland* (Paris: Council of Europe Development Bank, Dec. 2018)
6. Amy Frearson, "The UK can 'learn lessons from school-building in Brazil' - Aberrant Architecture", *Dezeen*, Oct. 7, 2012.
7. Oliver Wainright, "Flatpack or flexible? Oscar Niemeyer's schools could have lessons for the UK", *Guardian*, Mar. 15, 2013.
8. A lecture by Simon Allford: Architecture and the Art of the Extra Ordinary, British School in Rome, Jul. 18, 2017.
9. Jennifer Singer, "The next generation of School Design", *School Design Together* (London: Routledge Publishing, Edited by Pamela Woolner, 2015)
10. Anja Wiegel, "A Cascade of Terraces: The Heights by BIG", *Detail online*, Nov. 25, 2019.
11. Karissa Rosenfield, "BIG Reveals Design for 'Cascading' Secondary School in Virginia", *Archdaily*, Jan. 08, 2016.
12. Stefan Schreiner and Michael Sagmeister, *PORR IMPLEMENTS PPP MODEL IN VIENNA*, Issue 174/2019.
13. Viennese Campus Model - Smart City Wien

卡尔韦博德·费勒德学校
Kalvebod Fælled School

Lundgaard & Tranberg Arkitekter

一座带给学生和社区积极、健康、快乐生活的圆形教学楼
A circular school building provides an active, healthy and happy life to both the students and the community

位于西阿迈厄岛的这所新建的公共学校为人们终身学习奠定了基础：这里社交活动丰富，体育活动和孩子们的各种活动都在此展开。

卡尔韦博德·费勒德学校位于哥本哈根奥雷斯塔德南部一片受保护的景观区中，位于牧场和城市之间，它就像日晷，记录时光流逝，记录太阳运动轨迹，记录孩童从幼儿变成青少年。

这所学校是一个开放且热闹的社区。放学后，它就向公众开放。学生家长和当地社区居民可以使用体育馆和教室来进行锻炼、上课和举办社交活动。

建筑师的设计理念是充分利用场地，既不封闭，也不设围栏，尽力使活跃的社交场所与附近独特的景观和谐相融，重点是实现日常活动和学习合二为一。卡尔韦博德·费勒德学校为儿童和成人提供了坚实而鼓舞人心的学习和休闲场所。

在自然界中，"圆度"通常是因围绕一个中心集中或扩散引起的，就像将尘雾和气体压缩进一个天体中，就像石头扔进池塘产生的圆形涟漪。位于卡尔韦博德·费勒德学校建筑中心的体育馆活力四射，它的能量辐射到了所有的五个楼层。

从体育馆开始，一系列空间呈辐射状分布。首先，由混凝土柱子支撑的中庭贯穿建筑所有楼层，让所有楼层都能清晰地俯瞰体育馆。其次，构成楼板的环状结构随之出现，延伸到宽敞的公共区域，而在靠近立面的教室一侧空间变得密集，这里是进行集中课堂教学的地方。

学校一层向外面的景观和周围的城市延伸出去，因此，受保护的景观和室外操场之间和缓地融为一体。这种和谐的感觉与磨光发亮的室内混凝土楼板也相互呼应，就好像卡尔韦博德·费勒德学校的景观脱掉了绿色的外衣，露出下面明亮的盐田，中间还会偶尔露出埋在地下的岩石。

Vestamager's new public school creates the framework for life-long learning: here a social vision of sports and children leading active lives finds its shape.

Like a chronograph tracking the passing of the year, the path of the sun and children's journeys from early childhood into adolescence, Kalvebod Fælled School is situated in the protected landscape of Ørestad South in Copenhagen, between the grazing cattle and the city.

The school is an open, bustling community. When the school

day ends for the students, for others it begins; parents and local community residents use the gymnasium and classrooms for sports, courses and social events.

The ambition has been to fully seize the site without fences or barriers, while softly balancing the unique landscape and the active social community set to unfold here. Focus is on the everyday integration of movement and learning. Kalvebod Fælled School provides solid, inspiring learning and leisure conditions for children and adults alike.

In nature, circularity typically arises from concentration or dispersion around a center. Hazes of dust and gases compact into astronomical bodies or circular ripples extend from a pebble thrown into a still pond. At Kalvebod Fælled School, the gymnasium at the core of the building extends its energy to all five floors.

From the gymnasium a series of spaces radiate. Firstly, atria carried by exposed concrete columns that run the entire height of the building provide clear views of the gymnasium from all floors. Subsequently, the rings making up the floors ensue, spreading into generous common areas while densifying into classrooms closer to the facade, for focused classroom work.

On ground floor level the school reaches out towards the landscape and the adjacent city. The result is a gentle integration between the protected landscape and the outdoor playground, a fusion also echoed in the sparkling, polished concrete interior floors, as if the landscape of Kalvebod Fælled had shed its green coat to display the bright salt meadow below, gently speckled with embedded rocks.

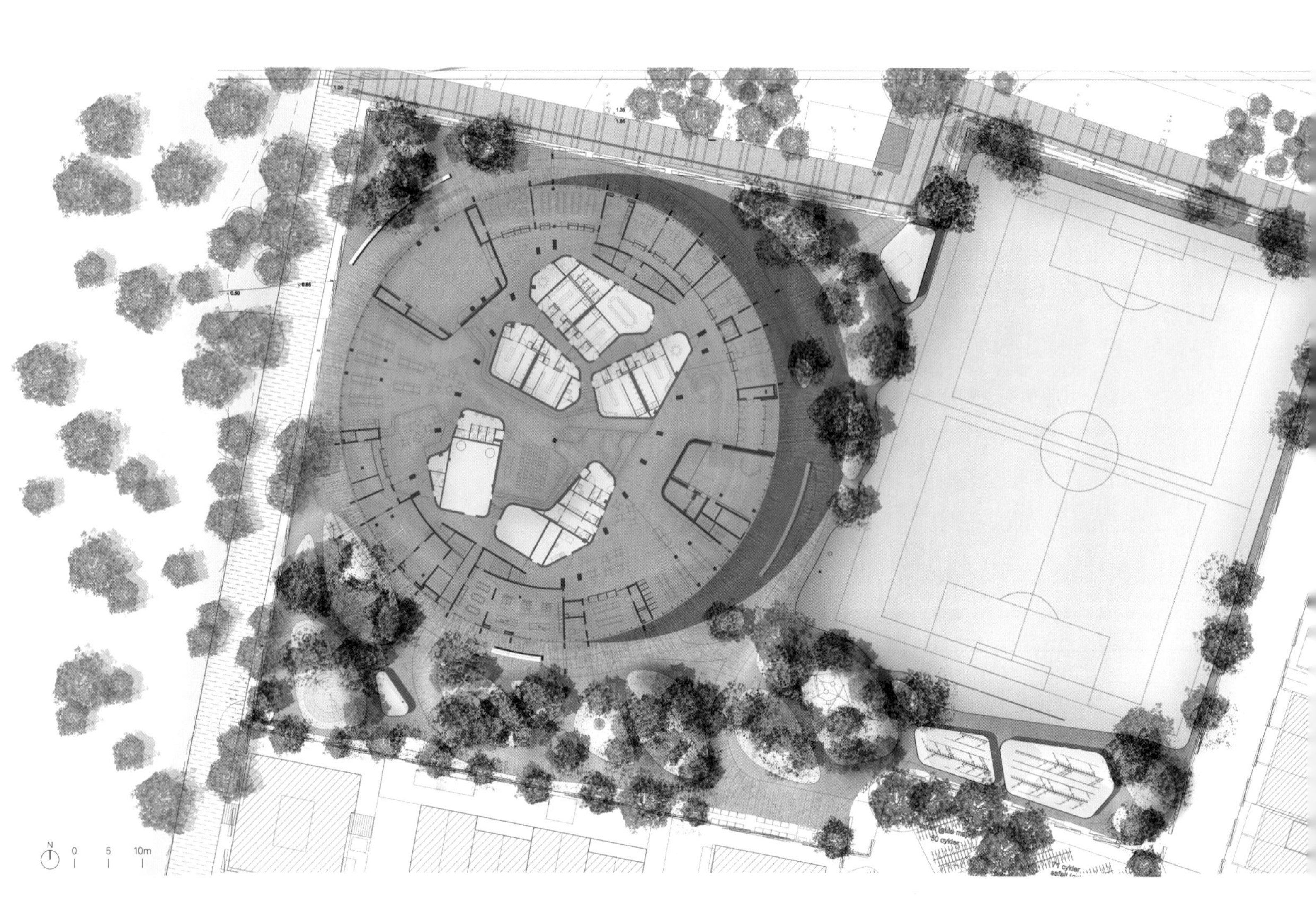

项目名称：Kalvebod Fælled School / 地点：Else Alfelts Vej 2, 2300 Copenhagen S, Denmark / 事务所：Lundgaard & Tranberg Arkitekter A/S / 施工与消防：Jørgen Nielsen Rådgivende Ingeniører A/S / 技术工程公司：Dansk Energi Management & Esbensen A/S / 声学设计：Gade & Mortensen Akustik A/S / 景观设计：BOGL / 客户：The Municipality of Copenhagen / Københavns Ejendomme / 建筑面积：11,560m² / 能源规范：Low Energy 2015, 38 KwH/m²/year / 项目总造价：USD 58 mill. / 设计竞赛时间：2014 / 施工时间：2016—2018 / 摄影师：©AndersSuneBerg (courtesy of the architect) - p.16~17, p.20~21, p.28, p.30 upper, p.33; ©TorbenEskerod (courtesy of the architect) - p.22, p.23, p.24~25, p.26, p.27, p.30 lower, p.31, p.34~35

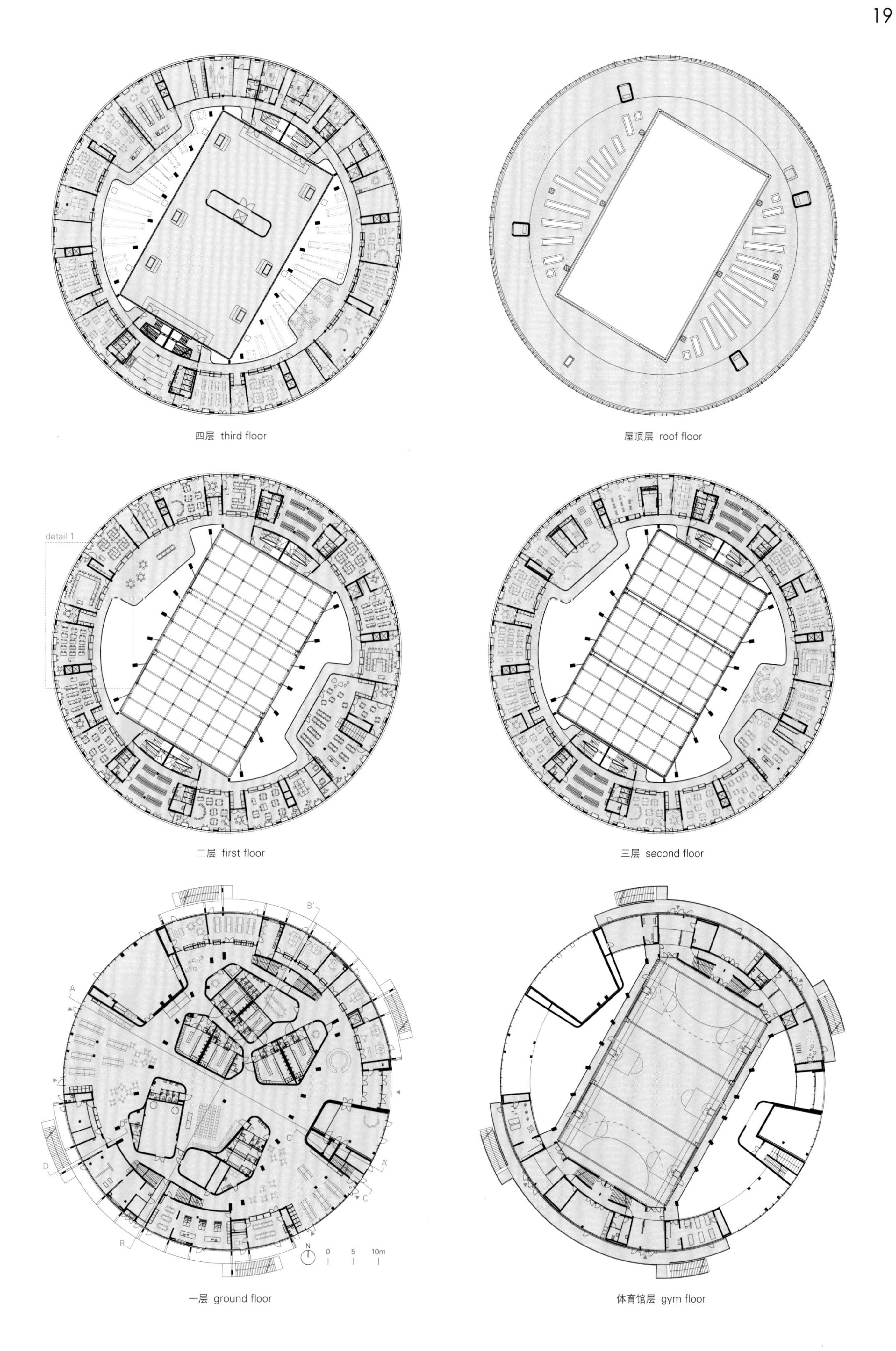

四层 third floor

屋顶层 roof floor

二层 first floor

三层 second floor

一层 ground floor

体育馆层 gym floor

3.22

Auditorium
Kantine
Lille sal
Omklædning

1
0

4
3
2
1
0

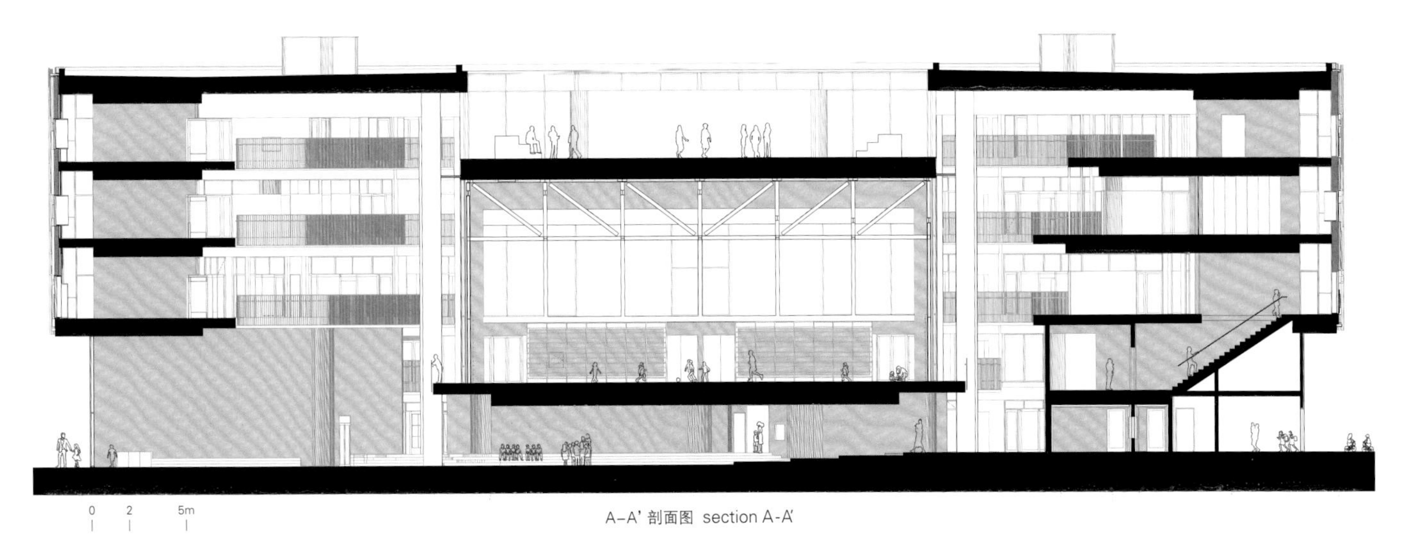

A-A' 剖面图 section A-A'

B-B' 剖面图 section B-B'

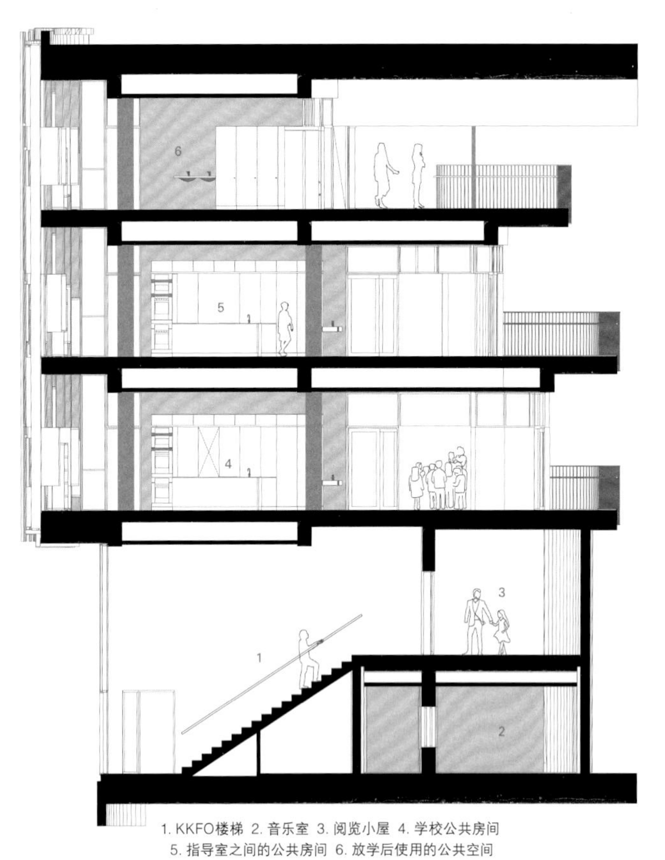

1. KKFO楼梯 2. 音乐室 3. 阅览小屋 4. 学校公共房间
5. 指导室之间的公共房间 6. 放学后使用的公共空间
1. KKFO staircase 2. music room 3. read cave 4. common room-schooling
5. common room-between Instruction 6. common room-leaving school after
C-C' 剖面图 section C-C'

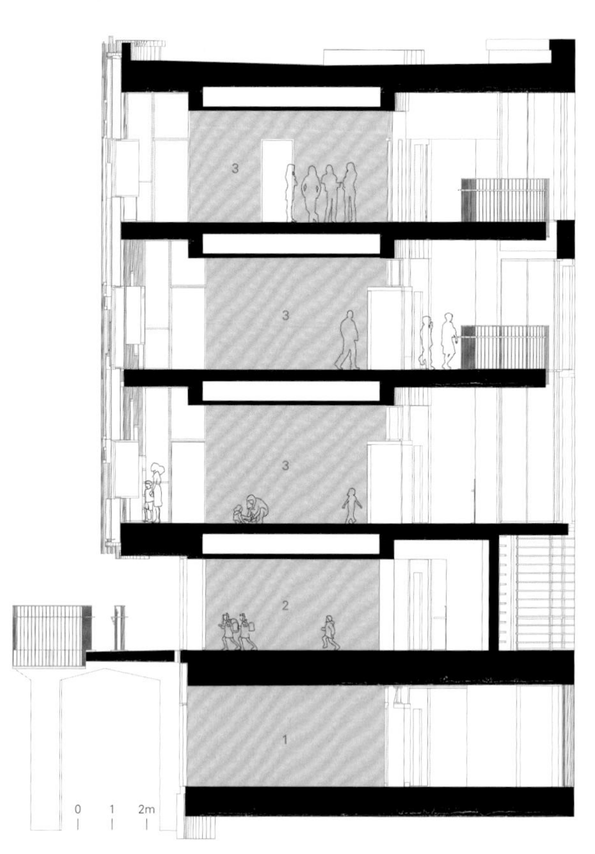

1. 共享厨房 2. 体育馆出口 3. 教室
1. shared kitchen 2. exit from the gym 3. classroom
D-D' 剖面图 section D-D'

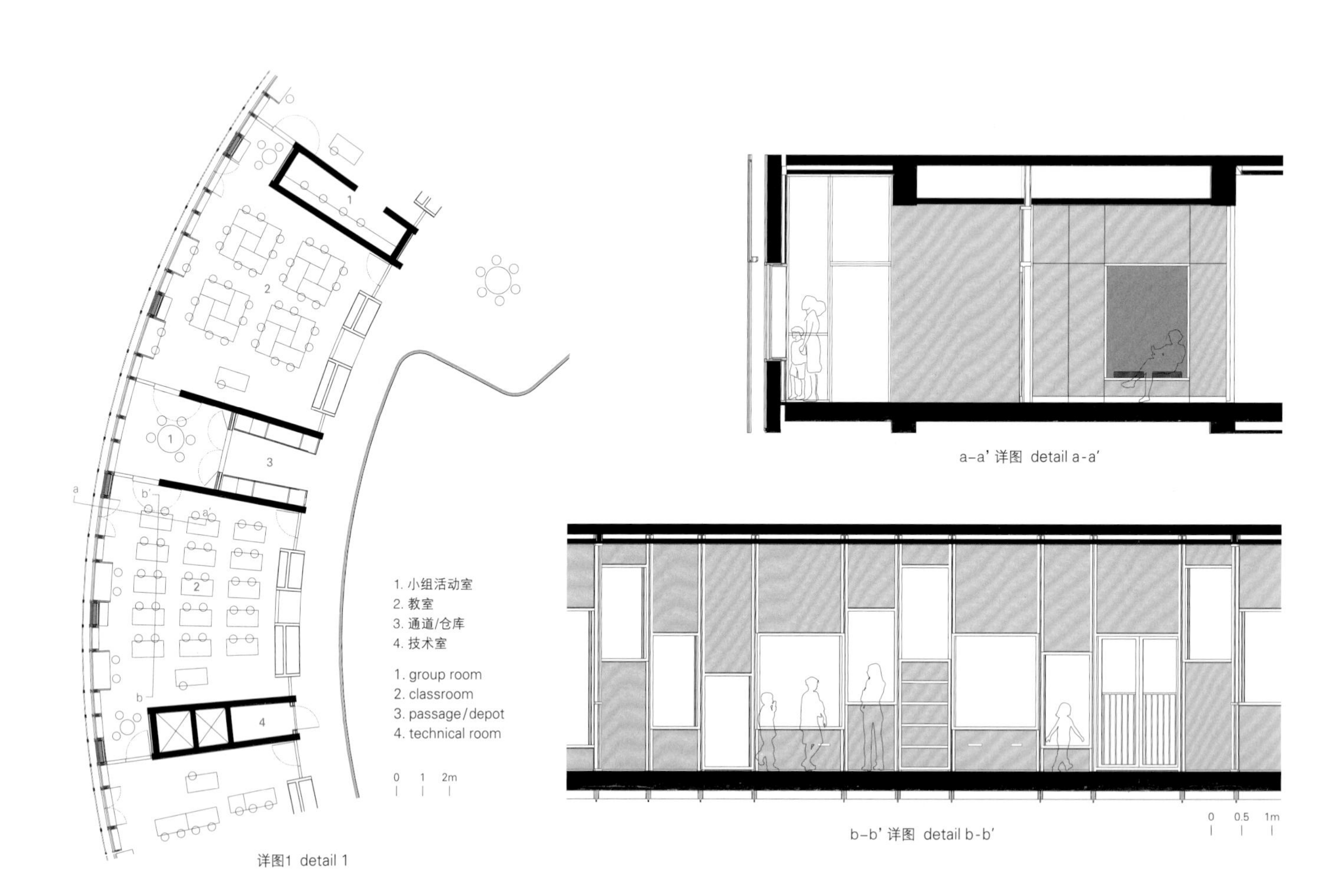

a-a' 详图 detail a-a'

b-b' 详图 detail b-b'

详图1 detail 1

4
3
2
1

红岭实验小学
Hongling Experimental Primary School

O-office Architects

红岭实验小学，一所高密度的中国南方小学

Hongling Experimental Primary School, a high-density primary school in the south part of China

红岭实验小学位于中国广东深圳福田区的一座山上，除了学校场地西侧的一小块孤零零的地方，山体已经几乎消失了。其余场地在采石工作逐步取消后，被平整为城市开发用地。红岭实验小学的建设用地约为10 000m²，原规划为24个班的小学，后因缺口巨大而增容至36个班，现建筑面积约为原规划的两倍。四周对地铁线路和道路的避让以及建筑规范对日照和建筑间距的规定，使得建筑设计面临诸多限制。

因此，校园垂直方向的设计策略变得至关重要。建筑师在红岭实验小学的设计中，尽力把建筑总高度控制在24m以下，鼓励水平方向的互动，在建筑和景观设计中尽量满足儿童的生理和心理特点。

教学楼几乎完全覆盖了可建设用地，建筑被分成东西两边高低不同的两个半区，从总平面图上看是两个镜像的E字形结构相互连接。西半区利用教室之间所必需的间距创造出两个曲线形边界的"山谷"庭院。庭院下沉至地下一层，与景色优美的斜坡花园结合在一起，为地下一层的文体设施和食堂空间争取了充足的采光和自然的通风。下沉的庭院通过南侧庭院的缓坡和北侧庭院的露天阶梯剧场，与上方自然起伏的一层相连。

200m环形跑道和运动场位于建筑东半区的四层，与西侧主教学楼的三层相连。运动场下方是可容纳300人的礼堂，再下面是一个半户外游泳池。西半区上面几层是课外教室和教师办公室，屋顶是学校的园艺农场。

学习单元——传统上被称为教室，是学校学生学习和交流的基本空间单元。建筑师针对深圳的亚热带气候，运用了成对组合的鼓形平面学习单元设计，避免阻隔自然通风。每层12间教室被分成三排，呈6个单元对排列。每个单元对组合都可以通过可移动的分隔墙连接或分开。与传统的长方形教室相比，鼓形平面展现出更大的灵活性与自由度，更有利于开展多样化的教学模式。学习单元富有韵律的折线与庭院的曲线为小朋友们创造了一个富有活力的半户外活动场地。

两座钢结构花园连接桥连接起两个"山谷"庭院，在庭院空间的上方增加了一份独特的观赏和游戏体验。这些庭院、动态布置的水平楼板、疏松的单元组织以及有机的绿化均是针对高密度的城市和亚热带的南方气候而采取的设计策略，是一种全新的空间探索模式。

Hongling Experimental Primary School (HEPS) stands on a hill, in Futian District, Shenzhen, Guangdong, China. The hill has almost disappeared, except for a small lonely part to the west of the school site. The rest of the terrain had been leveled into urban development land after the gradual withdrawal of quarrying operations. The site of HEPS, originally intended as a 24-class school, is about 100 by 100 meters. Its current capacity has been increased to 36 classes due to increased demand, with a total floor area double that of the original planning. The site is constrained on all sides by a subway line, a road and building codes relating to daylight and spacing of buildings.

Therefore, the design strategy for the vertical orientation became crucial; the architects made concerted efforts to control the building height within 24 meters, encouraging horizontal interactions, and trying to respond to the physical and psychological characteristics of children within the architectural and landscape designs.

The school building, divided into two halves of different heights on the east and west, almost fully covers the buildable land. The general plan is two interlinking mirrored E-shapes. The west half uses the obligatory spacing between

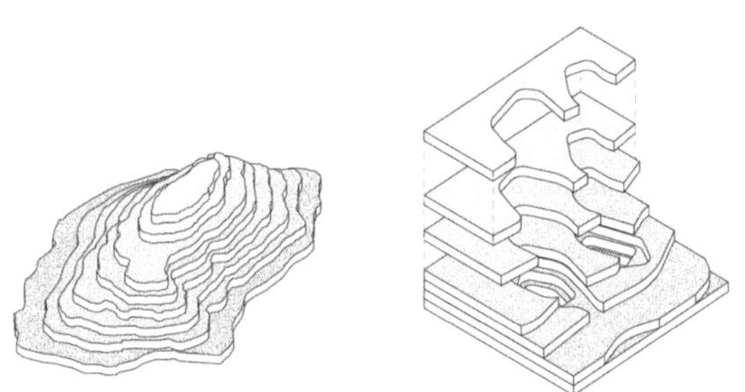

剩余的花岗岩山坡和新教学楼设计概念
the remaining granite hill and the conception of the new school building

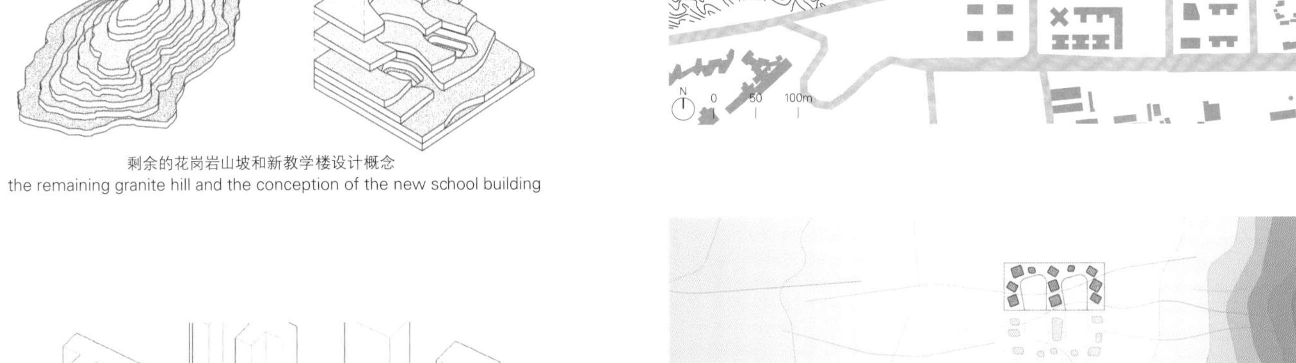

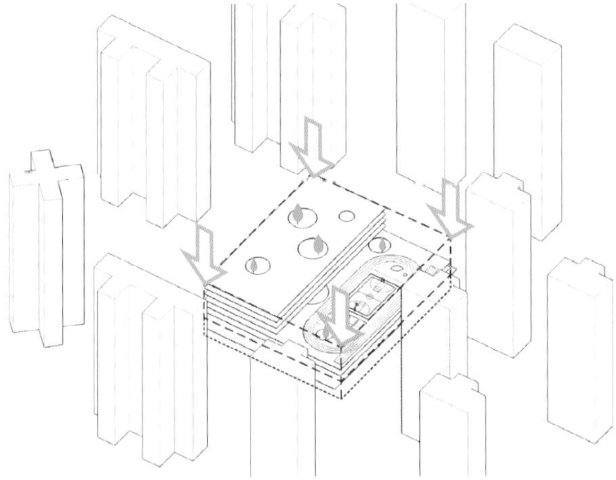

学校低层设计策略
the low-rise strategy of the school

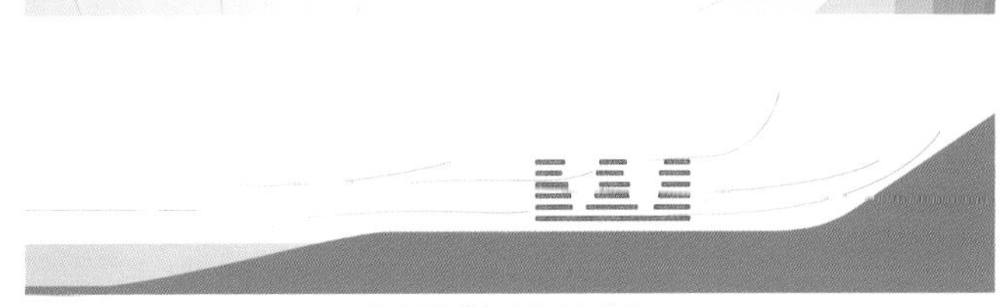

针对亚热带气候的空间策略
the spatial strategy in response to the subtropical climate

地形公园上方的山谷庭院
valley courtyards above the landform park

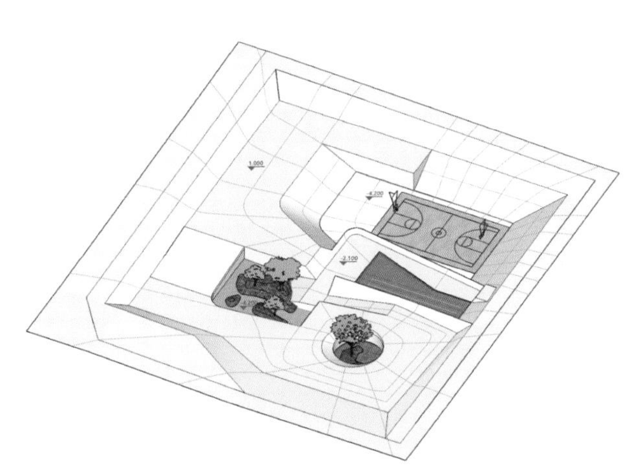

地形公园的形成
formation of landform park

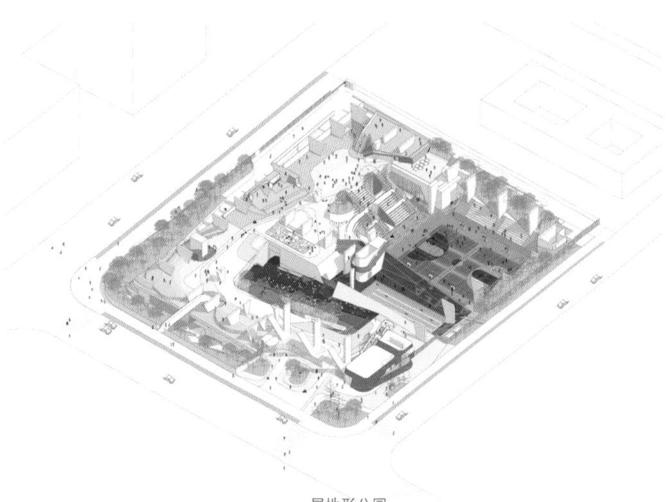

一层地形公园
ground landform park

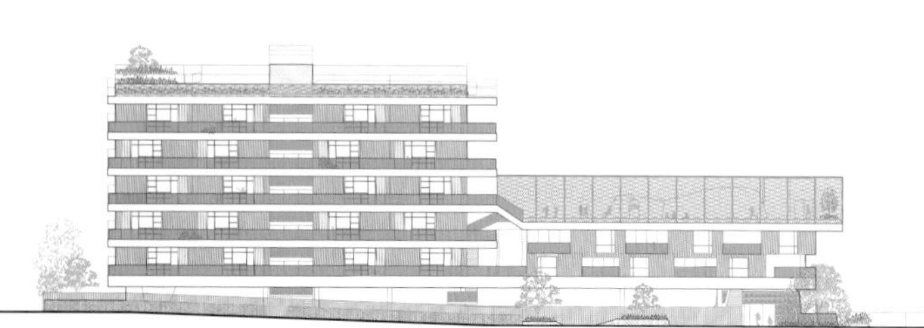

南立面 south elevation

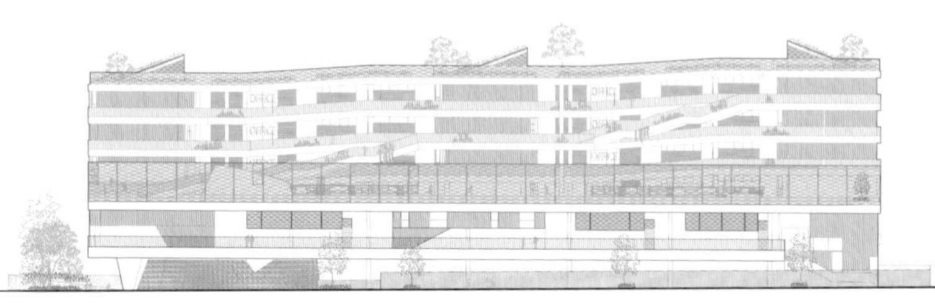

东立面 east elevation

北立面 north elevation

西立面 west elevation

classroom rows to create two "valley" courtyards with curved boundaries. The courtyards sink to an underground level, combining with a landscaped sloped garden, to become an underground cultural and sports facility, and a fully illuminated and naturally ventilated canteen space. The sunken courtyard is connected to the naturally undulating ground floor overhead through a gentle slope in the south courtyard and an open-air stepped theater in the north courtyard.

The 200-meter circular runway and sports ground are placed on the third floor of the eastern half, linking to the second level of the main teaching building on the western side. Underneath the stadium is a 300-seat auditorium above a semi-outdoor swimming pool. The upmost floors of the west half are extracurricular classrooms and teacher's offices, while the roof is a horticultural farm.

The learning unit, traditionally the classroom, is the basic spatial cell for school students to learn and communicate. Here, the architects conceived separated pairs of drum-shaped learning units, to avoid obstructing fluent ventilation in the subtropical Shenzhen climate. Every 12 classrooms are divided into three rows and arranged in six pairs. Each unit-pair combination can be joined or separated via a moveable division wall. The drum-shaped plan shows greater flexibility and freedom when compared to the traditional rectangular classroom and is more conducive to a variety of learning and teaching patterns. The rhythmic folding curved outline of the learning units and the curved edge of the courtyards shape a dynamic semi-outdoor venue for the children.

Two steel garden bridges connect the two "valley" courtyards, adding unique viewing and playing experiences over the space. These courtyards, the dynamic horizontal slabs, loose cellular fabric, and organic greening are the design strategies to respond to high-density urban conditions and subtropical southern climates, developing a new spatial productive paradigm.

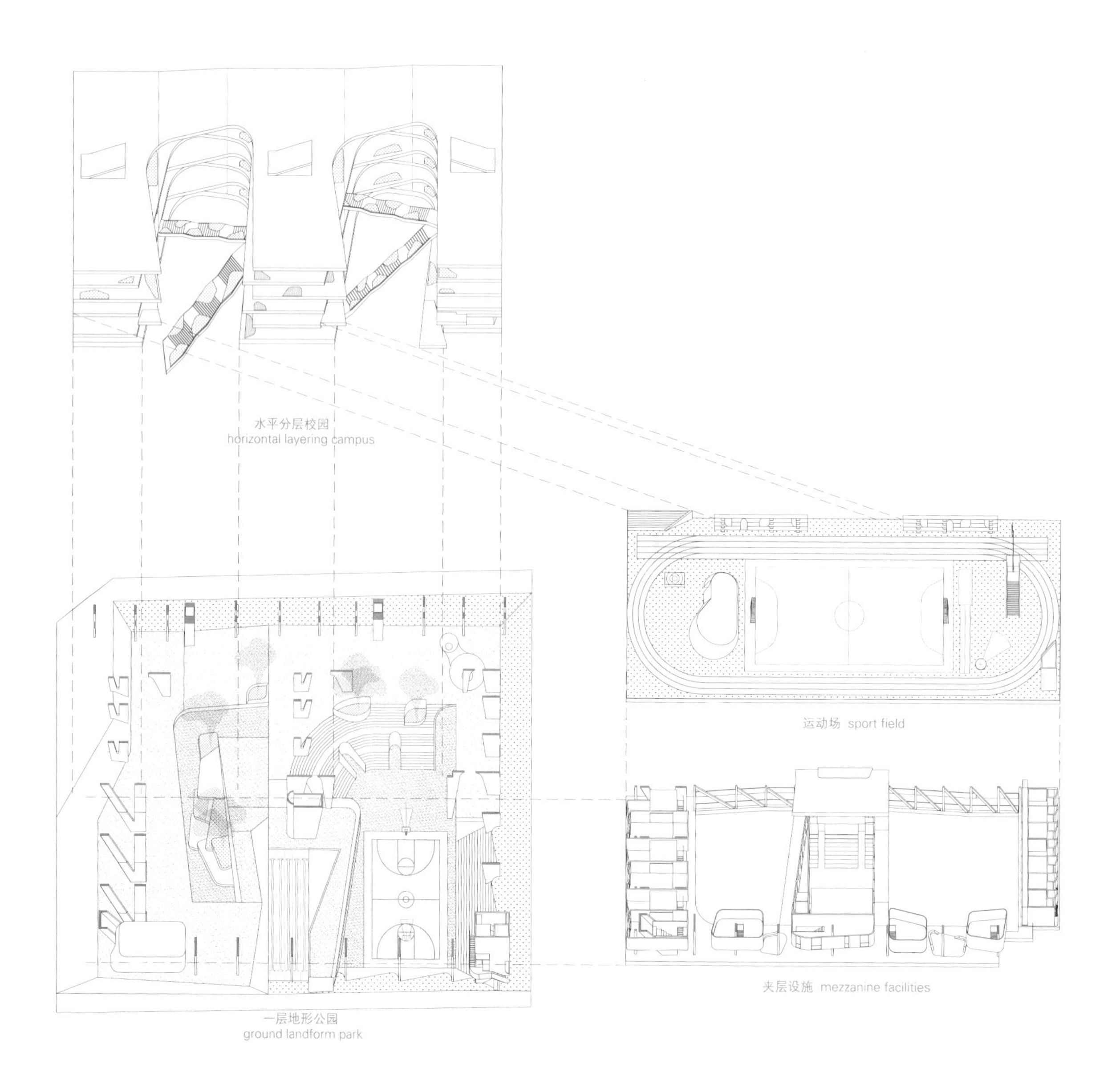

STUDENT
CANTEEN
学生食堂

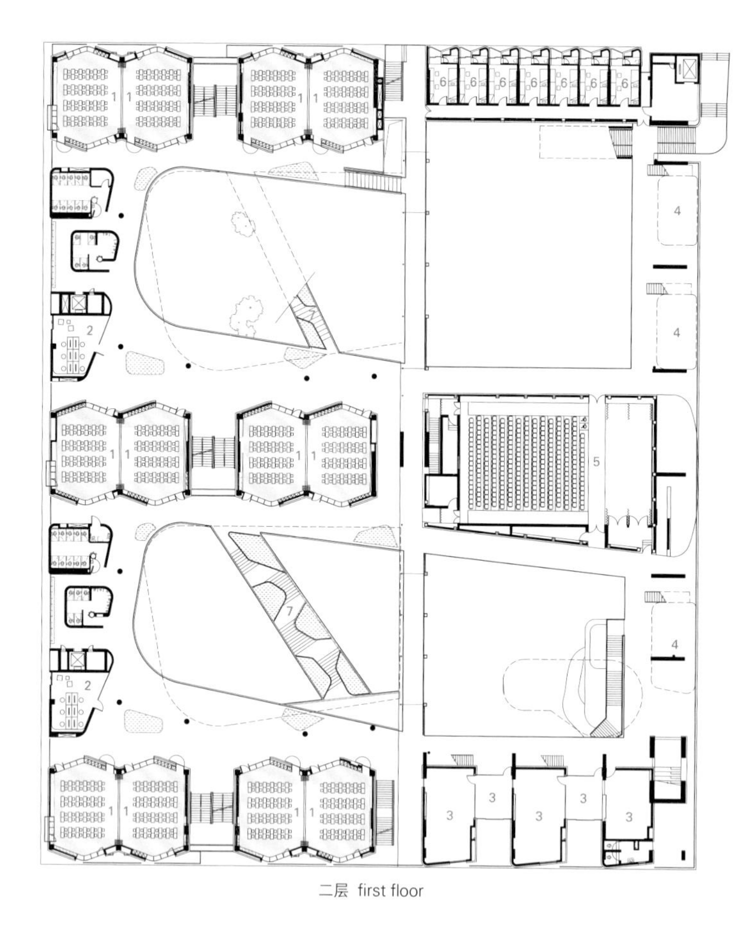

二层 first floor

1. 学习单元
2. 教师办公室
3. 行政管理办公室
4. 学生兴趣小组空间
5. 礼堂（284座）
6. 教师宿舍
7. 绿色连接桥

1. learning unit
2. teacher's office
3. administrative office
4. student interest group space
5. auditorium (284 seats)
6. teacher's dormitory
7. green bridge

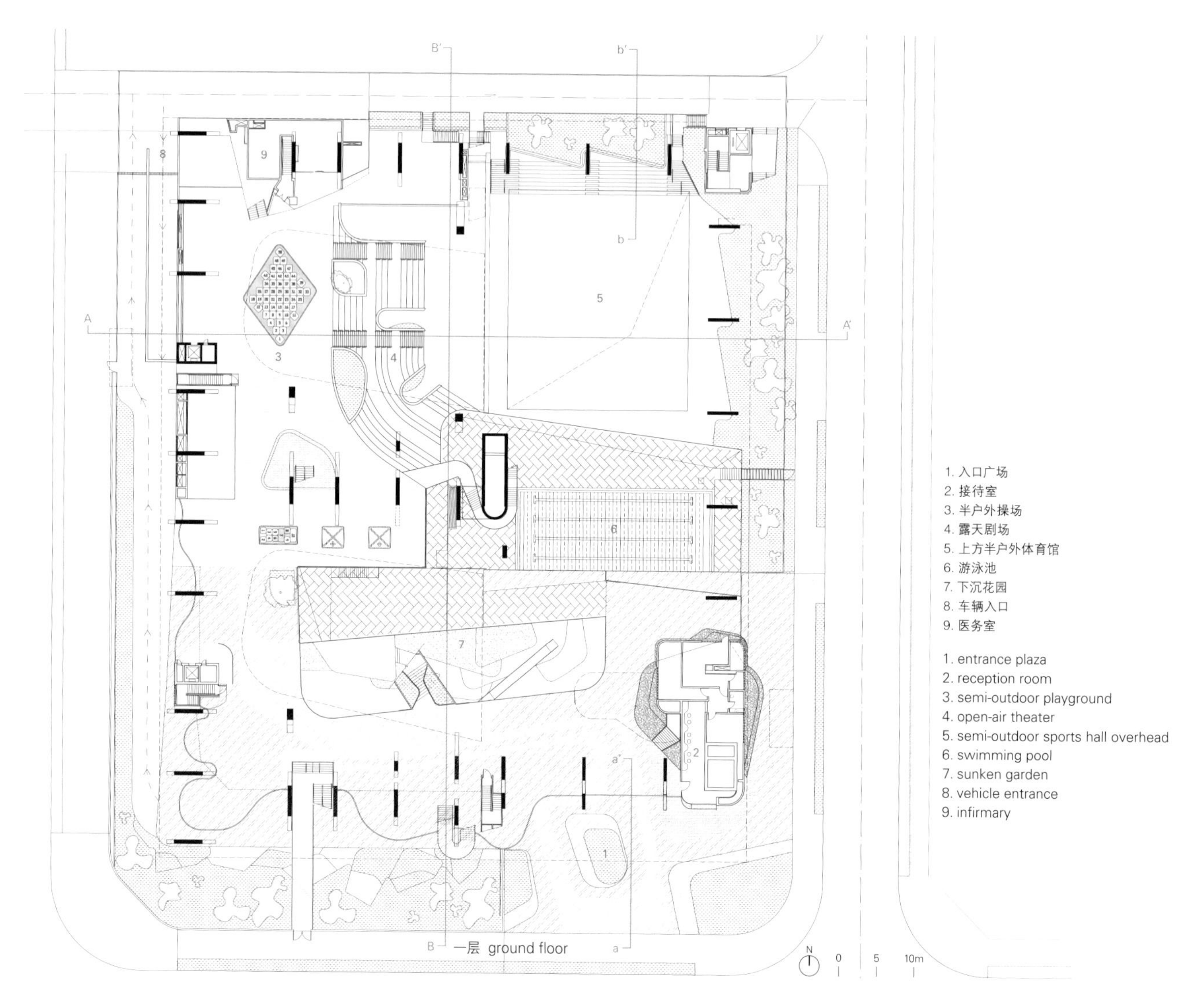

一层 ground floor

1. 入口广场
2. 接待室
3. 半户外操场
4. 露天剧场
5. 上方半户外体育馆
6. 游泳池
7. 下沉花园
8. 车辆入口
9. 医务室

1. entrance plaza
2. reception room
3. semi-outdoor playground
4. open-air theater
5. semi-outdoor sports hall overhead
6. swimming pool
7. sunken garden
8. vehicle entrance
9. infirmary

项目名称：Hongling Experimental Primary School
地点：Northeast of the intersection of Qiaoxiang 4th Road and Antuoshan 2nd Road, Futian District, Shenzhen, China
事务所：O-office Architects
总建筑师：He Jianxiang & Jiang Ying / 项目建筑师：Dong Jingyu, Chen Xiaolin
设计团队：Wu Yifei, Zhang Wanyi, Wang Yue, Huang Chengqiang, Zeng Wei, Cai Lehuan, Peng Weisen, He Zhenzhong
施工项目管理：Shenzhen Vanke Real Estate Co., Ltd.-Zhao Siyi, Huang Xin, Wang Chuan (site project management)
结构顾问：RBS Architectural Engineering Design Associates
结构与机电设计：CMAD Design Group
其他顾问：Neuco Building Facades Technology Co., Ltd. - Building facades design; TOP DESIGN - Landscape design; TheWhy art x design-VI Design; Shenzhen Guangyi Lighting Planning and Design Co., Ltd.-Lighting design; Shenzhen QianDian construction and Engineering Design Consulting Co., Ltd.-Structural overrun consultant
总建筑面积：33,721m² / 高度：24m / 竣工时间：2019
摄影师：
©Zhang Chao (courtesy of the architect) - p.36~37, p.38, p.40~41, p.43, p.44~45[left, right-upper], p.48~49, p.52[lower]
©Wu Siming (courtesy of the architect) - p.44~45[right-lower], p.50[bottom-left, bottom-right], p.52~53,
©Huang Chengqiang (courtesy of the architect) - p.52[upper]

1. 学习单元	1. learning unit
2. 教师办公室	2. teacher's office
3. 户外运动场	3. outdoor sport field
4. 绿色连接桥	4. green bridge
5. 年级聚集室	5. grade gathering room
6. 科学室	6. science room
7. 书法室	7. calligraphy room
8. 陶艺室	8. pottery room
9. 校园广播室	9. campus broadcasting room
10. 心理咨询室	10. counseling room
11. VR室	11. VR room
12. 语言实验室	12. language lab

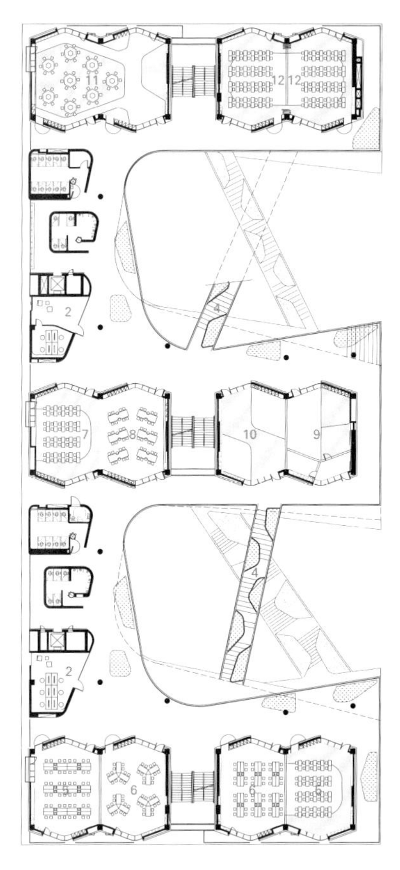
五层 fourth floor

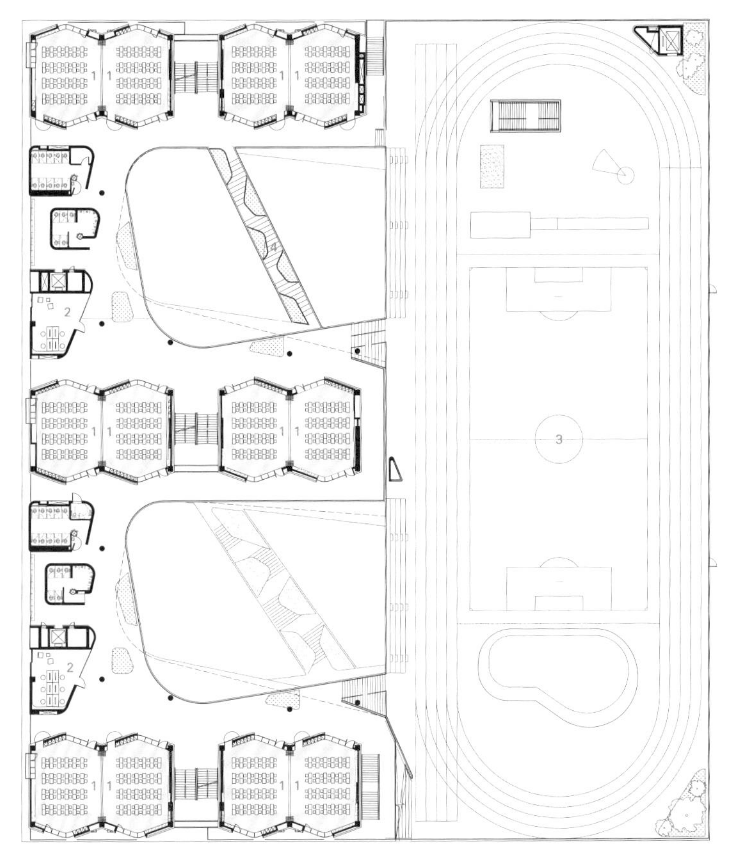
三层 second floor

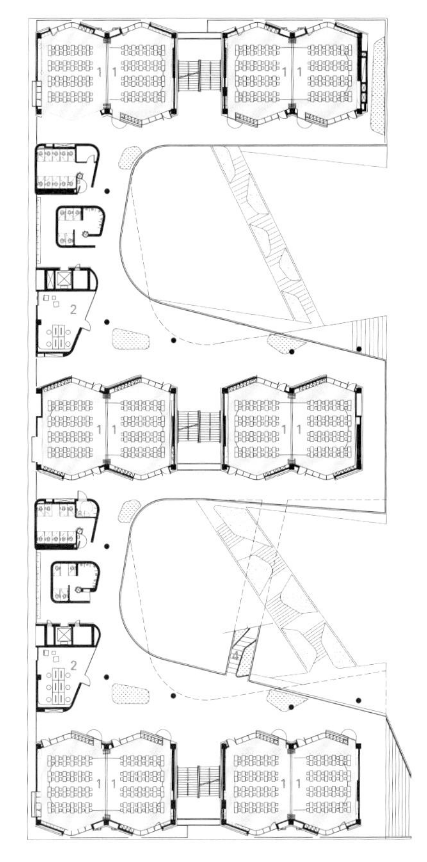
四层 third floor

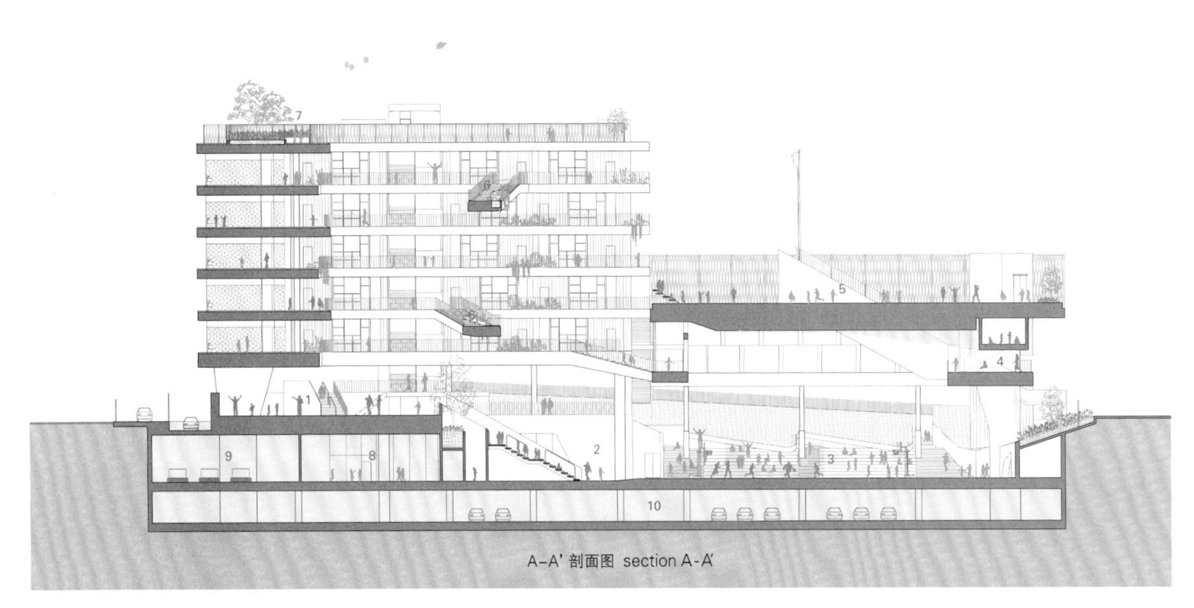

A–A' 剖面图 section A-A'

1. 地形公园
2. 露天剧场
3. 半户外体育馆
4. 学生兴趣小组空间
5. 户外运动场
6. 绿色连接桥
7. 屋顶农场
8. 架高活动空间
9. 技术室
10. 地下停车场

1. landform park
2. open-air theater
3. semi-outdoor sports hall
4. student interest group space
5. outdoor sport field
6. green bridge
7. roof farm
8. raised-up activity space
9. technical room
10. underground parking

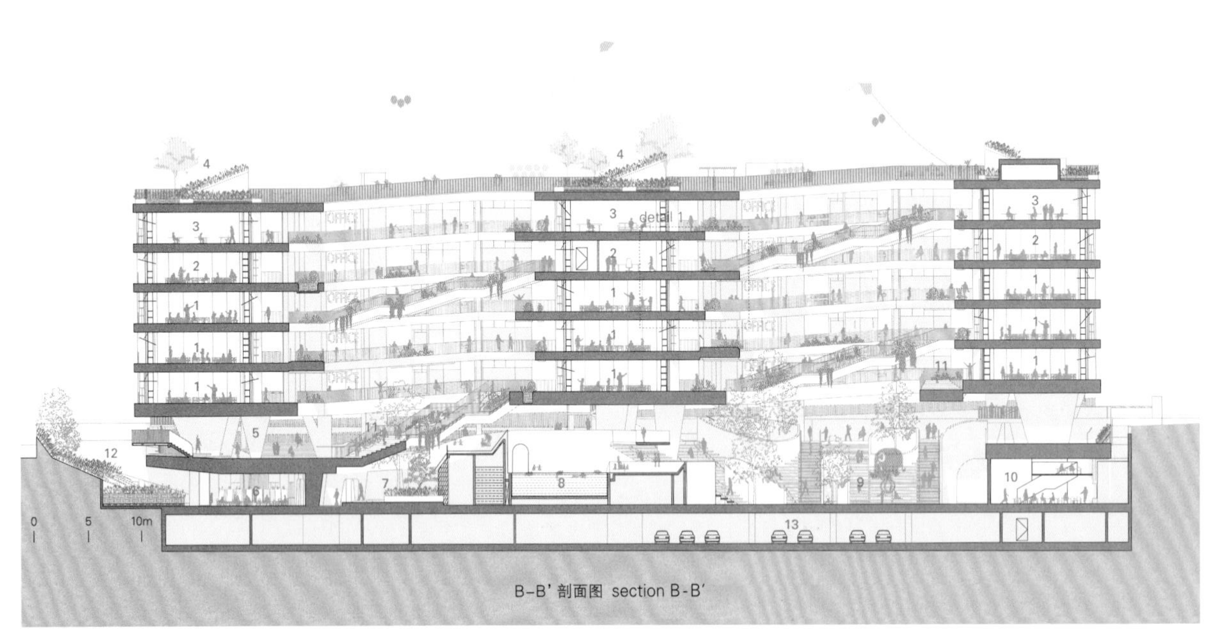

B–B' 剖面图 section B-B'

1. 学习单元
2. 课外学习空间
3. 行政管理办公室
4. 屋顶农场
5. 地形公园
6. 艺术学习空间
7. 艺术花园
8. 游泳池
9. 露天剧场
10. 食堂
11. 绿色连接桥
12. 侧面坡地花园
13. 地下停车场

1. learning unit
2. extracurricular learning space
3. administrative office
4. roof farm
5. landform park
6. art learning space
7. art garden
8. swimming pool
9. open-air theater
10. canteen
11. green bridge
12. side sloping garden
13. underground parking

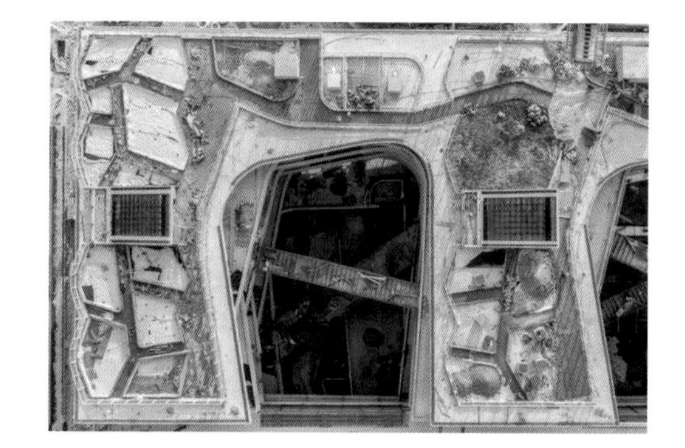

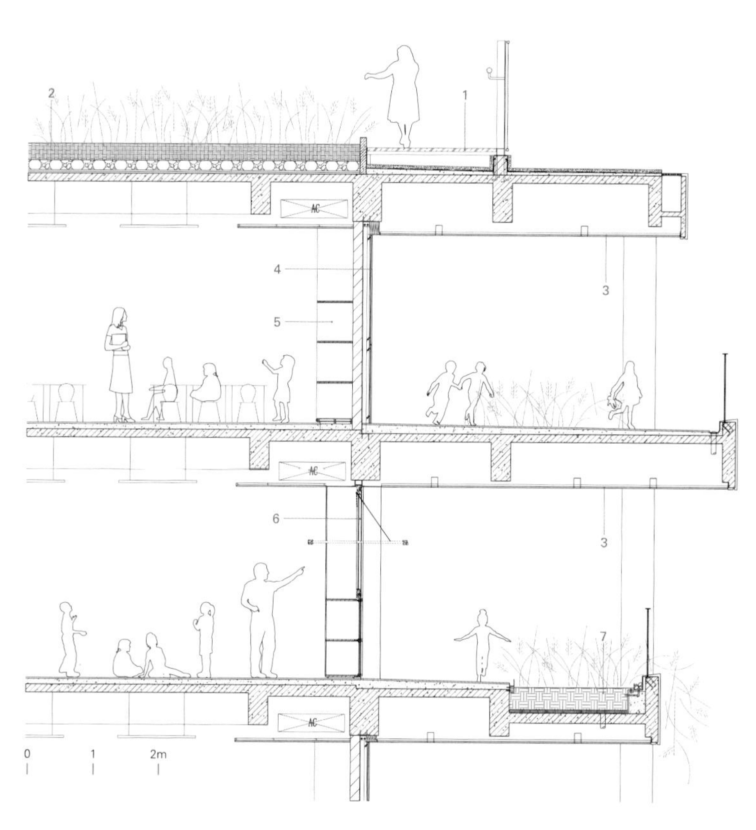

详图1 detail 1

1. refractory brick paving 2. roof farming system 3. fiber-reinforced panel 4. terracotta Baguette facade 5. steel storage rack 6. electric center-pivoted window 7. small planting outdoor garden 8. prefabricated rubber runway 9. perforated aluminum plate 10. steel-mesh fence 11. washed granitic floor 12. phenolic resin laminated panel 13. side garden 14. sloping garden system 15. silicone PU floor

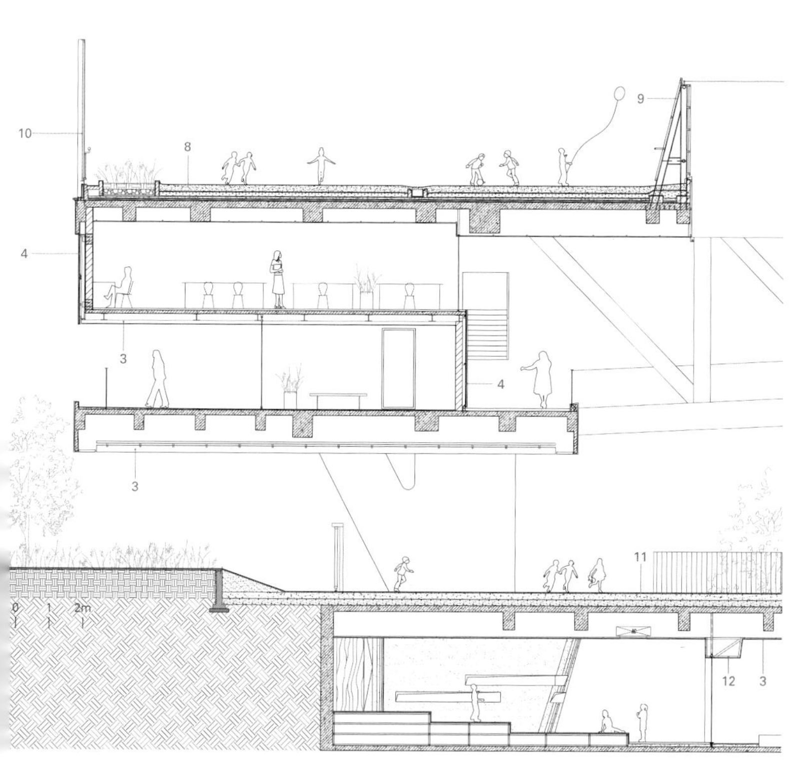

a-a' 详图 detail a-a'

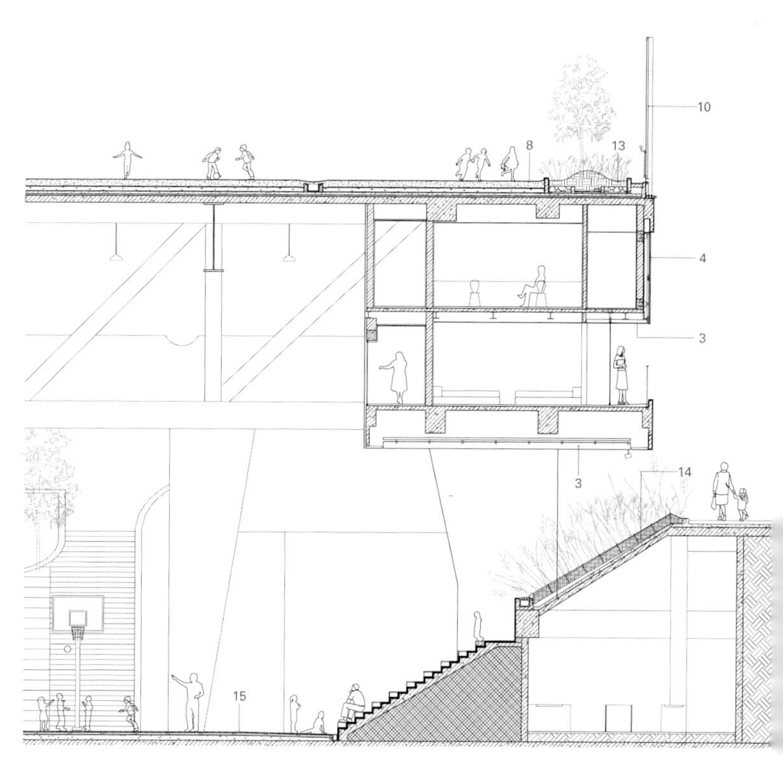

b-b' 详图 detail b-b'

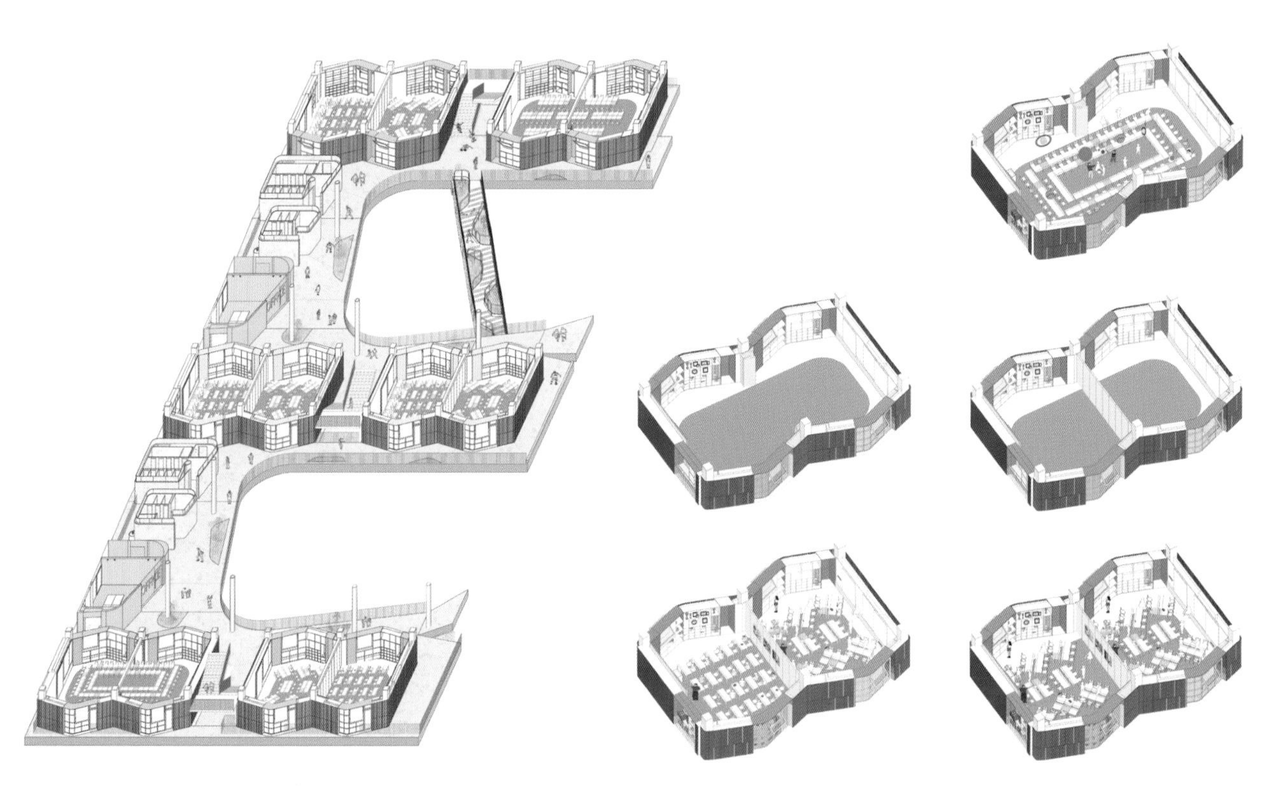

E字形楼板上6对学习单元
the 6 pairs of learning units on the E-shape floor plate

成对的鼓形学习单元为主动学习过程提供了较高的灵活性
the pair drum-shape learning unit offering a high flexibility in interactive learning process

万花筒式的教育园区
The educational campus as a kaleidoscope

贝雷斯加斯教育园区是维也纳一个开发区发展的起点，这里将提供3000多间公寓、办公室、商店和休闲设施。该项目的目标是最大限度地发挥校园各部分之间以及校园与当地之间的空间协同效应，将新的教育园区定位为新教育实践实验室，类似于万花筒。

教育园区的建筑体量呈南北走向，贯穿整个场地，形成了三个不同的、对比鲜明的室外区域：第一个是西面超过3000m²的向公众开放的V形前庭；第二个是一个种植面积超过10 000m²的丰富种植区，是园区里一个共享、开放的绿色空间；第三个区域是运动场和多功能硬地球场。

大型建筑群被组织成一系列易于识别和管理的学习集群，每个学习集群都有一个独立的入口，在城市环境中很容易识别。

学习集群（BildungsbereichBiBer）是围绕中央公共区域组织的一组自主教室和辅助空间。这样的设计方式可以将相对较大数量的44间教室按更易于管理的空间单元组合在一起，从而加强小组和班级之间的合作。教学人员的合并工作区（小组室）能确保这些学习集群成为自主的组织单元。

该建筑一层包含所有普通空间，二层和三层为幼儿园和小学，四层为新的中学。

从建筑形态上看，该建筑的设计有点像金字塔，在一层呈矩形，而到了四层，建筑体量紧密地连接在一起。这就在上部楼层形成了约2000m²的宽大的、相互关联的露台。有些露台有植被丰富的绿化岛，有效地将花园扩展成一个三维空间。

贝雷斯加斯教育园区
Education Campus Berresgasse

PSLA Architekten

这种整体的组织原则使建筑的上层具有很强的结构分区。就规模而言，它与相邻的两种类型（独栋住宅和沿街建筑）有很大的区别。鉴于该项目场地的规模，对于一个教育园区来说，这令该项目拥有了独特的特点和尺寸。

该建筑可以看作是一系列类型的共生体：联排建筑、板式结构、桥梁结构和大厅结构的混合体。本案这所学校是第一个建成的如此规模的学校项目，它采用辐射状的风车一样的布置方式，有意识地替代了教室典型的正交形式布局，这样的布局方式能增加空间彼此之间的相互关系，也能改善空间与城市环境的关系。

各个学习集群中庭的中央公共区域都可以有多种使用方式。不同区域专为不同年龄段的人互动使用，也可供同一个学习集群中不同年龄的人共同使用，以增加他们的集体观念。

这些中庭的高度超过7m，在窗帘、座椅长廊、活动隔断和较低的天花板高度的帮助下，最大限度地扩大了视觉和空间的联系，同时也为人们提供了足够的休息和认真工作的机会。

在一些公共区域可以看到周围地区的景色，而另一些公共区域则形成内部阳台，面向中庭。每一个星形学习集群都能接收到来自各个方向的光照。

建筑体量的连接方式、"跳跃式"窗户的布置、横竖交替的落叶松木板条立面以及幼儿园的游戏室设计，这些都让师生们有一种宾至如归的感觉。

Berresgasse Educational Campus is the starting point for a developing district in Vienna, which will become home to over 3,000 apartments, offices, shops and leisure facilities. The objective of the project is to maximize the spatial synergies between the different parts of the campus, and between the campus and the local area, positioning the new educational campus as a laboratory for new educational practices – similar to a kaleidoscope.

The north-south oriented volume of the educational campus meanders across the site, creating three different and contrasting external areas: a publicly accessible, V-shaped forecourt of over 3,000m^2 to the west, a richly planted area of over 10,000m^2 of shared, open, green space for the campus, and a third area on Berresgasse dominated by the sports hall and multipurpose hard court.

The large building complex is organized into a sequence of easily recognizable and manageably scaled learning clusters, each of which has a separate entrance and is clearly legible in its urban context.

A learning cluster (Bildungsbereich or BiBer) is an autonomous group of classrooms and ancillary spaces organized around a central common area. This approach enables the relatively large number of 44 classrooms to be grouped together in more manageable spatial units that strengthen cooperation between groups and classes. Incorporating working areas (team rooms) for teaching staff ensures that these BiBers become autonomous organizational units.

The building is divided into a ground floor – containing all the general spaces; the kindergarten and primary school on the first and second floors; and the new middle school,

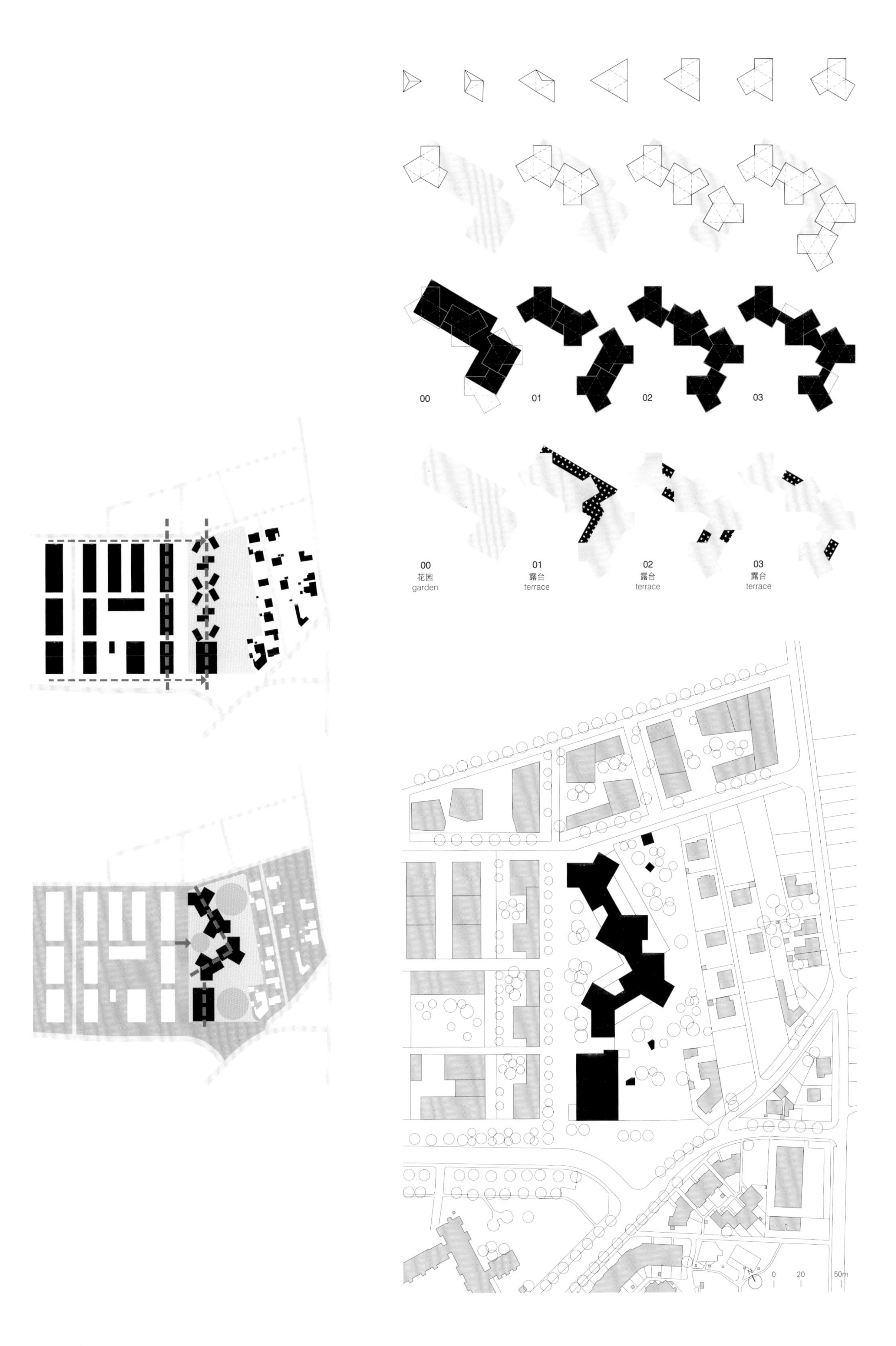
00
01
02
03
00
花园
garden
01
露台
terrace
02
露台
terrace
03
露台
terrace
0
20
50m

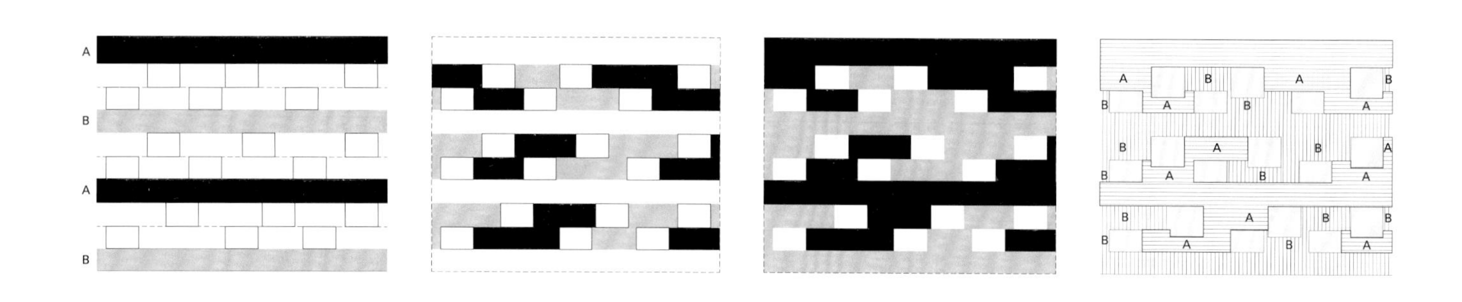
A
B
A
B

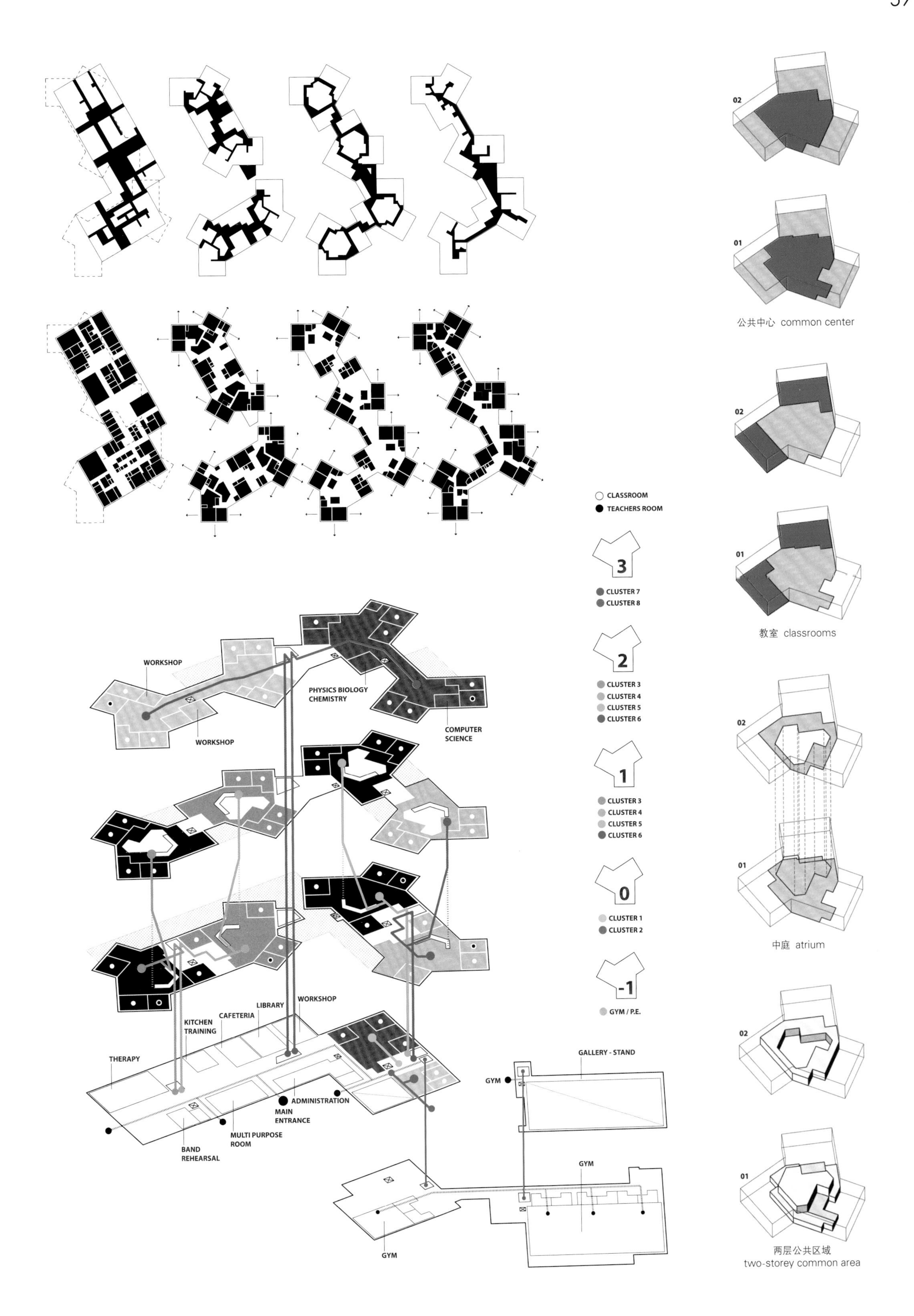
02
01
公共中心 common center
02
01
教室 classrooms
02
01
中庭 atrium
02
01
两层公共区域
two-storey common area
CLASSROOM
TEACHERS ROOM
3
CLUSTER 7
CLUSTER 8
2
CLUSTER 3
CLUSTER 4
CLUSTER 5
CLUSTER 6
1
CLUSTER 3
CLUSTER 4
CLUSTER 5
CLUSTER 6
0
CLUSTER 1
CLUSTER 2
-1
GYM / P.E.
WORKSHOP
WORKSHOP
PHYSICS BIOLOGY
CHEMISTRY
COMPUTER
SCIENCE
THERAPY
KITCHEN
TRAINING
CAFETERIA
LIBRARY
WORKSHOP
ADMINISTRATION
MAIN
ENTRANCE
MULTI PURPOSE
ROOM
BAND
REHEARSAL
GALLERY - STAND
GYM
GYM
GYM

一层 ground floor

二层 first floor

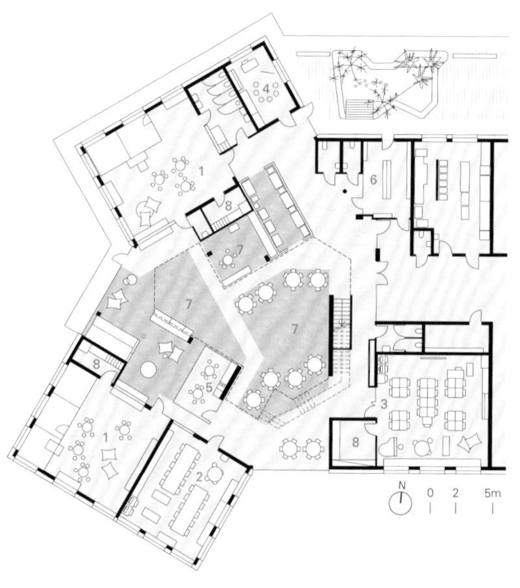

1. 教室 2. 教职员工办公室 3. 学前班/音乐室 4. 音乐教育室
5. 小厨房 6. 衣帽间 7. 公共区域 8. 储藏室
1. classroom 2. staff room 3. preschool/music 4. music education
5. kitchenette 6. cloak area 7. common area 8. storage room
二层——学习集群 first floor_cluster

三层 second floor

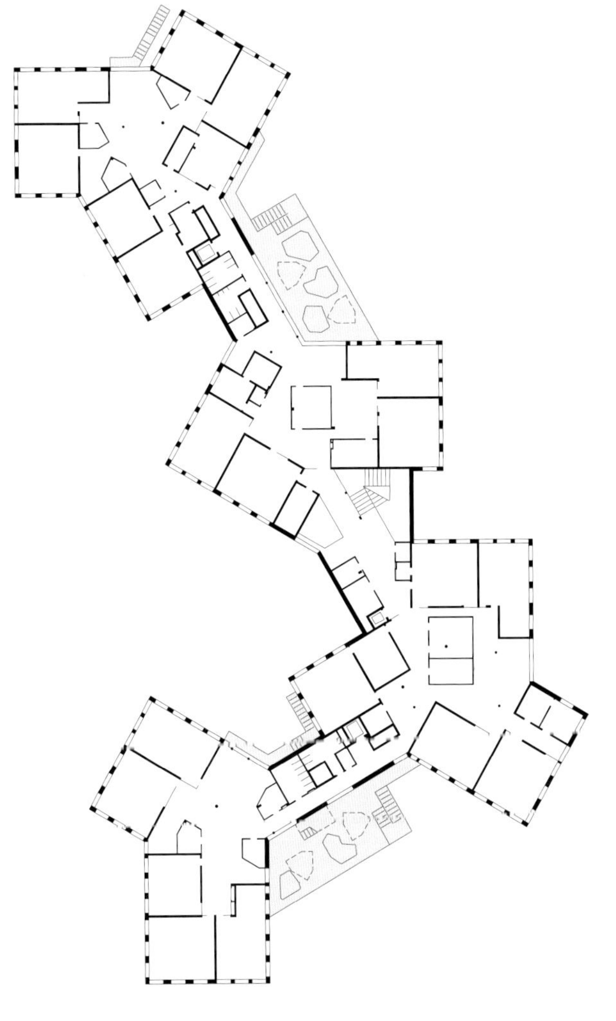

四层 third floor

1. 教室 2. 后加房间 3. 衣帽间 4. 公共区域 5. 储藏室
1. classroom 2. add-on room 3. cloak area
4. common area 5. storage room
三层——学习集群 second floor_cluster

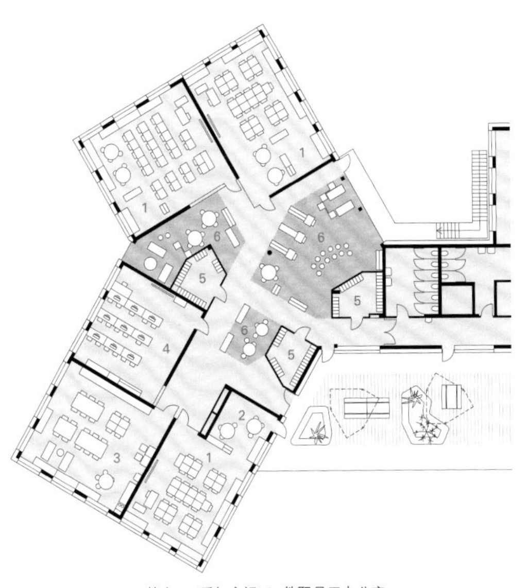

1. 教室 2. 后加房间 3. 教职员工办公室
4. 计算机科学室 5. 衣帽间 6. 公共区域
1. classroom 2. add-on room 3. staff room
4. computer science 5. cloak area 6. common area
四层——学习集群 third floor_cluster

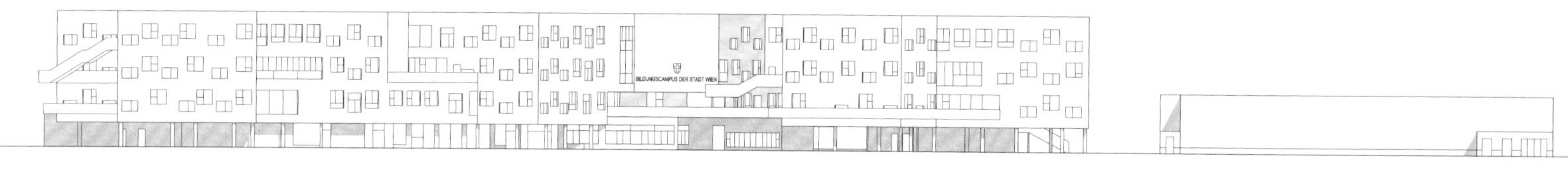

西立面 west elevation

东立面 east elevation

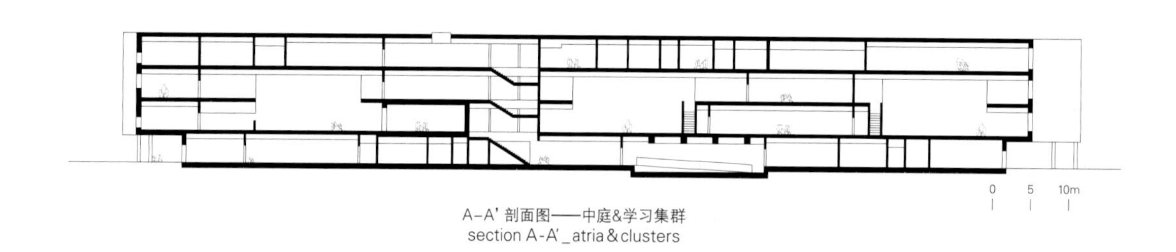

A–A' 剖面图——中庭&学习集群
section A-A'_atria&clusters

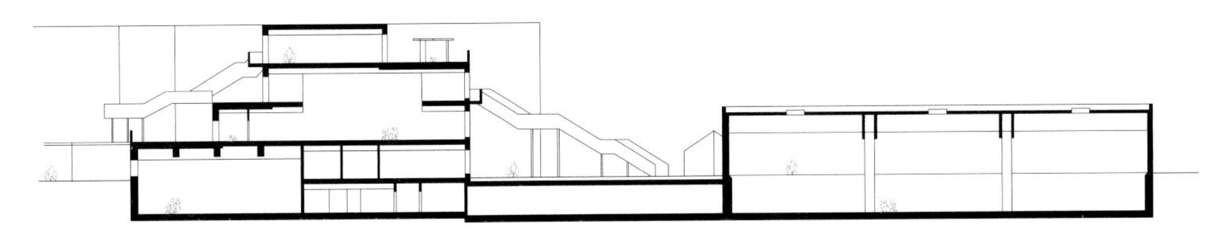

B–B' 剖面图——体育馆&中庭集群
section B-B'_gym&atrium clusters

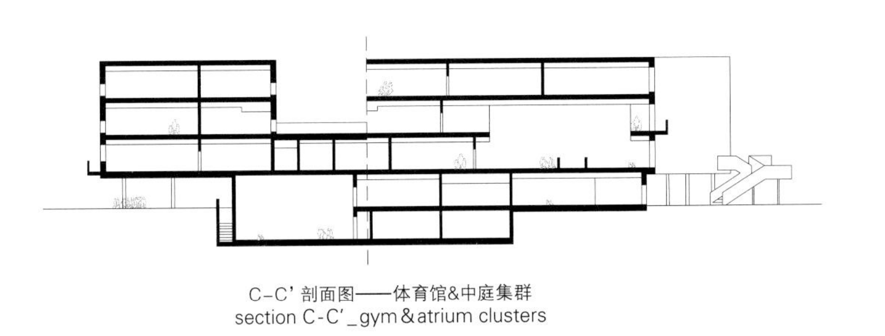

C–C' 剖面图——体育馆&中庭集群
section C-C'_gym&atrium clusters

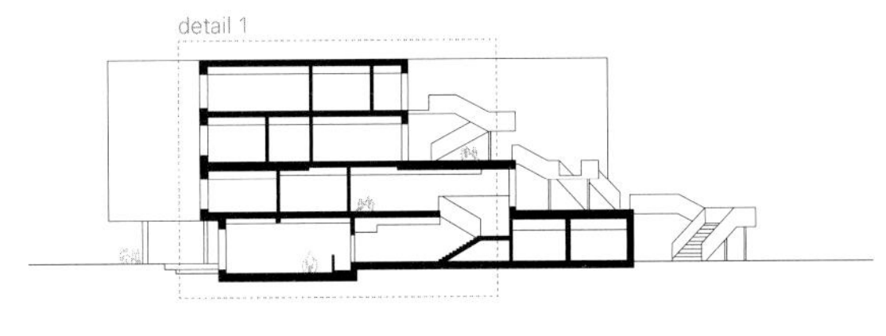

D–D' 剖面图——多功能室&露台
section D-D'_multi-purpose room&terrace

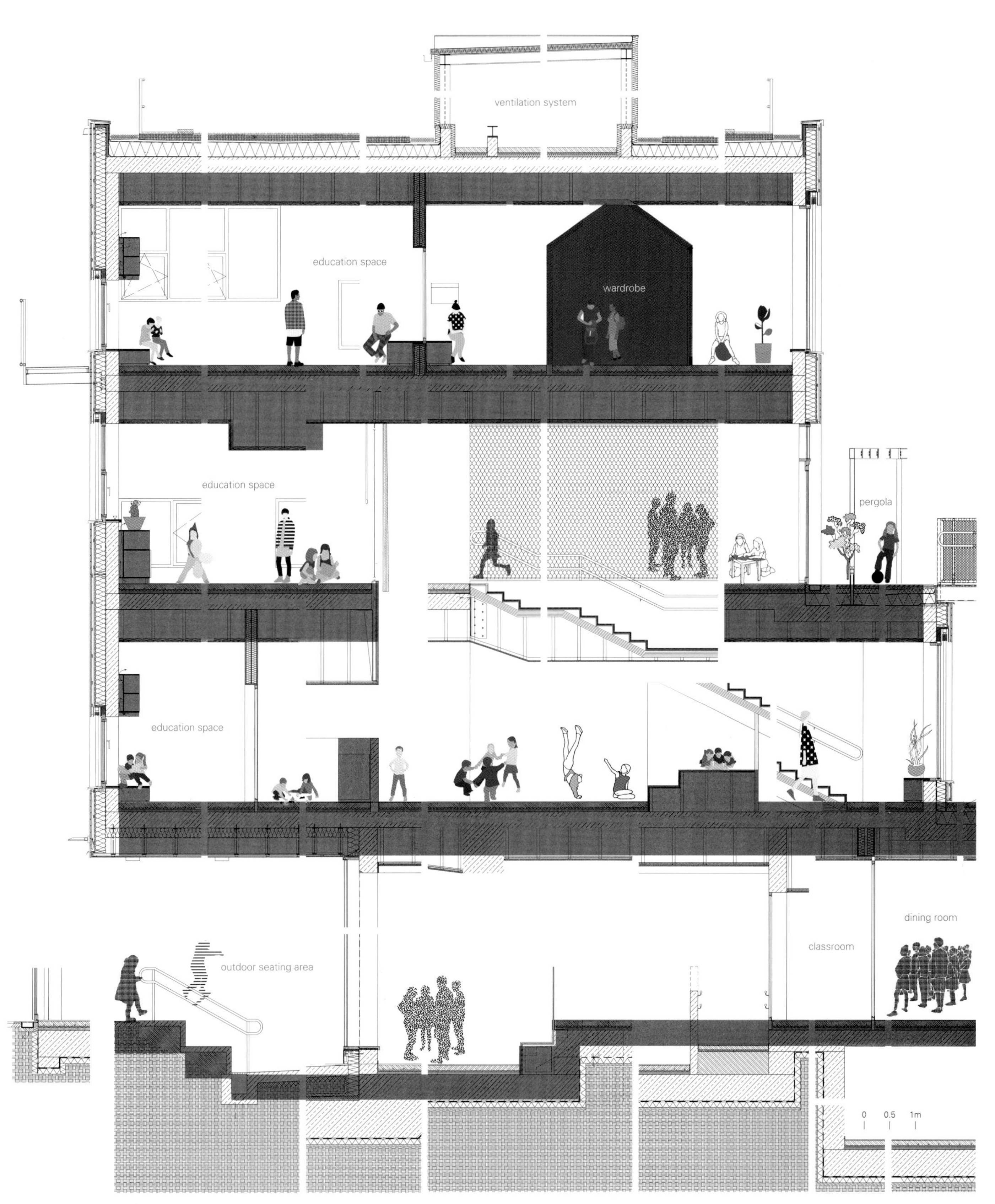
ventilation system
education space
wardrobe
education space
pergola
education space
outdoor seating area
classroom
dining room
0
0.5
1m

详图1 detail 1

located on the third floor.

In morphological terms, the building undergoes a pyramid-like reduction from the rectangular ground floor to the strongly articulated volume at the fourth level. This results in a total of around 2,000m^2 of generous, interconnected terraces at all upper levels, some of which contain islands of rich greenery that effectively expand the garden into a third dimension.

This overall organizing principle leads to a strong constructional differentiation between the building's upper stories, which are very different from the two neighboring typologies – detached houses and street front blocks – in terms of scale. Given the size of the site, this gives the project an independent character and dimension appropriate for an educational campus.

The building can be seen as a symbiosis of a range of typologies: a hybrid of a terraced, slab, bridge and hall structure. This is the first completed school project of this scale that has consciously replaced the orthogonal organization of the classrooms with a radial, windmill-like configuration, a kaleidoscope-like expression, that enhances the relationship of the spaces with each other and with the urban context.

The central common areas, in the atria of the individual BiBers, can be used in a wide variety of ways. Their different zones permit age-specific interaction but also encourage a sense of community within a "BiBer family" that transcends the wide age-difference.

With a height of over seven meters, these atria maximize visual and spatial connections while also generating sufficient opportunities for retreat and concentrated work, with the help of curtains, seating galleries, mobile partitions and reduced room heights.

Some common areas offer views out to the surrounding area while others, which form internal balconies, are oriented towards the atria. Each of the star-shaped learning clusters is illuminated by daylight from every direction.

The articulation of the volume, the position of the "jumping" windows, the alternating horizontal and vertical larchwood slatted facade, and the design of the playhouses in the kindergarten ensure that the building complex is a lively home for pupils and teachers.

项目名称：Education Campus Berresgasse / 地点：Scheedgasse 2, 1220 Vienna, Austria
建筑设计、家具、总体规划：PSLA Architekten ZT GMBH
项目团队：Lilli Pschill, Ali Seghatoleslami, Aiste Ambrazeviciute, Christopher Ghouse, Andreas Hradil, Marc Werner, Anna Barbieri, Andreas Metzler, Roland Basista
顾问：EGKK Landschaftsarchitektur (landscape architecture), FCP Fritsch, Chiari und Partner ZT GmbH (structural design, building physics, fire safety), rhm GmbH (HVAC), TB Eipeldauer (electrical planning) / 总承包商、施工规划：Porr Bau GmbH/ Porr Design & Engineering / 客户：City of Vienna / 用途：education /
用地面积：19 070m^2 / 建筑面积：7 190m^2 / 总楼面面积：19 281m^2
建筑造价：€ 38.9 Mio / 竞赛时间：2015 / 施工时间：2017—2019
摄影师：©Lukas Schaller (courtesy of the architect)

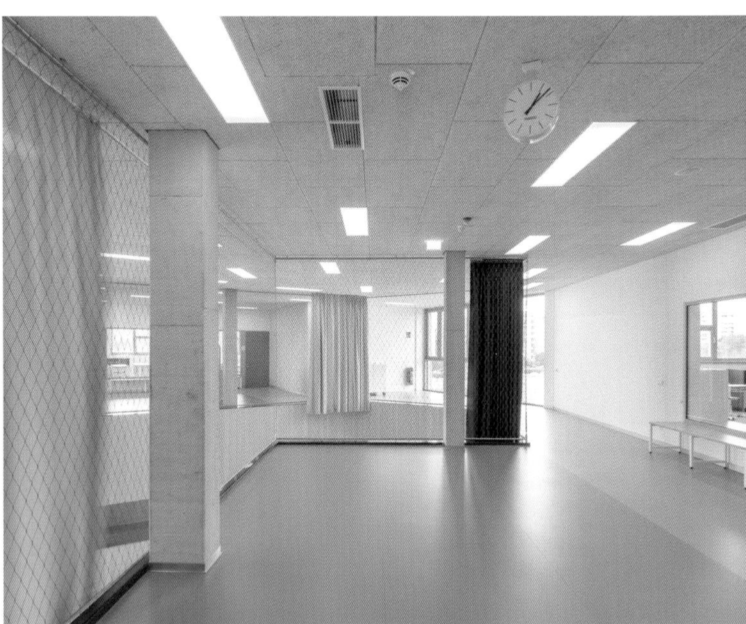

海茨学校
The Heights School

BIG

从中心轴呈扇形向外展开的层叠教室——BIG设计的海茨学校教学楼

Cascading classrooms fan outwards from a central axis in BIG's Heights School's building

海茨学校位于阿灵顿罗斯林–巴尔斯顿走廊沿线，同时将现有的两所中学——HB Woodlawn和Eunice Kennedy Shriver容纳在一栋面积为16 700m²的建筑中，可容纳775名学生。这一设计支持学校信任与自主的价值理念，并能为有特殊教育需求的学生提供大量资源。

海茨学校教学楼的整体形态犹如倾泻而下的绿色梯田，从中心轴呈扇形展开，在密集的城市环境中成为一个垂直的社区。

海茨学校坐落在市区内一片较为紧凑的区域内，三面被道路环绕，还有一侧与罗斯林高地公园相连。建筑物由五个矩形结构以围绕中心固定点推进旋转的方式堆叠构成，保留了传统单层教学楼所拥有的集体感和空间效率。每一层的绿色露台都是教室空间的延伸，创造了一个室内外连通的学习景观，相较于传统的教学环境，它打造的是一片传递知识的绿洲。位于中央的旋转楼梯横穿建筑内部，连接每层露台，让学生可以轻松来到室外，并在社区和学校之间建立更紧密的联系。上层露台更适合作为安静的学习区，而宽敞的二层露台和1740m²的休闲场地也可作为全校和社区活动的场所。

从威尔逊大道的校门进入海茨学校，学生和教职员工们会先经过一个三层高的大厅，大厅内设有阶梯式座位，可供学生集会和公众集会使用。学校的许多公共空间，比如，含400个座位的礼堂、体育馆、图书馆、接待处和自助餐厅等，都与大厅直接相连。建筑师面向社区设计了轻松方便的进入方式，鼓励公众积极参与学校内举办的活动。这样的设计还营造了富有温度和亲和力的环境，同时增强了共享空间之间的视觉联系。学校内还设置了专供学生使用的空间，包括一间艺术工作室、科学和机器人实验室、音乐排练室和两个表演艺术剧院。教室的储物柜成为连接各空间的主要元素，围绕着包含电梯、楼梯和洗手间的垂直核心筒展开布局。当学生们从中央楼梯进入时，迎接他们的是颜色的渐变：每间教室的储物柜都有各自不同的颜色，因此，在发挥导向功能的同时，也带来了充满活力的社交氛围，这种氛围从一楼到顶楼都有。Shriver学校旨在为11至22岁的学生提供特殊教育，它占据两层楼，从一层可以直接进入。这里有专门用于支持APS功能性生活技能训练项目的空间，这些空间既有私密性，又方便进入，包括体育馆、庭

院、职业理疗室和感官室，这些空间都以帮助感官训练和恢复为目的。海茨学校的外观采用了优雅的白色釉面砖，将五个建筑体量统一起来，并突出以扇形扭转的教室储物柜之间形成的斜角，充分体现建筑的雕塑感、建筑内部的活力和活动。建筑师考虑到周围的社区和曾经的威尔逊学校，选用了向亚历山大古城的建筑表达敬意的建筑材料。

Located along Arlington's Rosslyn-Ballston corridor, The Heights merges two existing secondary schools – the HB Woodlawn Program and the Eunice Kennedy Shriver Program – into a new 16,700m² building which accommodates 775 students. The design supports the school's ethos of trust, self-governance and the provision of extensive resources for specialized educational needs.

The Heights School's building opens as a cascade of green terraces fanning from a central axis, forming a vertical community within a dense urban context.

Situated within a compact urban site bounded by roads on three sides and a portion of Rosslyn Highlands Park, The Heights is conceived as a stack of five rectangular floorplates that rotate around a fixed pivot point, maintaining the community feeling and spatial efficiencies of a one-story school. Green terraces above each floor become an extension of the classroom, creating an indoor-outdoor learning landscape – an educational oasis rather than a traditional school setting. A rotating central staircase cuts through the interior, connecting the terraces, allowing students to circulate outside and forge a stronger bond between the neighborhood and the school. While the upper terraces are more suitable for quiet study areas, the spacious first terrace and 1740 m² recreation field also serve as event venues for school-wide and neighborhood activities.

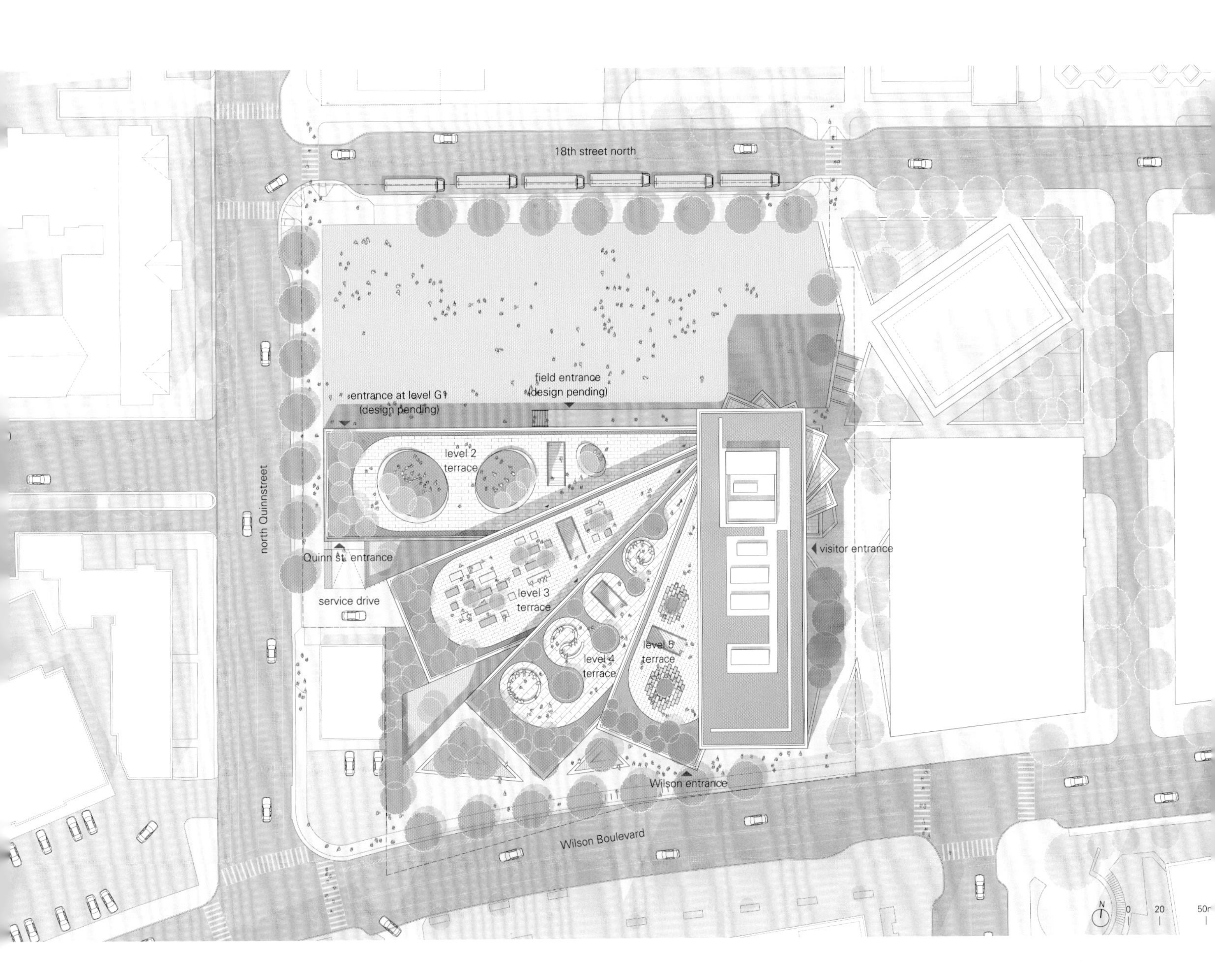

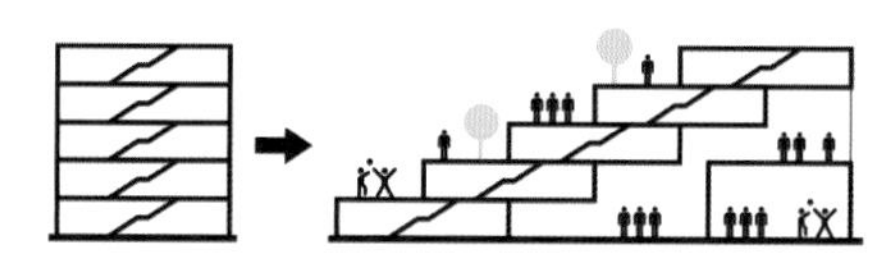

垂直学校 + 社区

VERTICAL SCHOOL + COMMUNITY

The small site requires the project to be designed across multiple levels. A key objective for the design was to maintain the feeling of a 1-story school building while still having a vertical school and the efficiencies afforded by it.

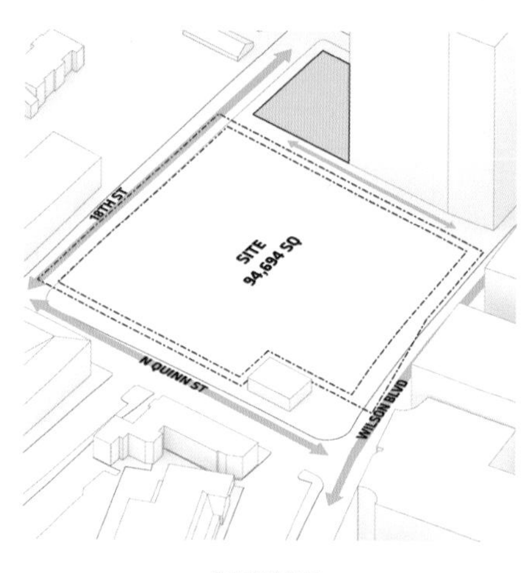

海茨学校场地

THE HEIGHTS BUILDING SITE

The compact urban site is bounded by roads on three sides with an existing 7-11 convenience store located on the busy corner of Wilson Boulevard and North Quinn Street. Along 18th Street, the site shares a common edge with Rosslyn Highlands Park.

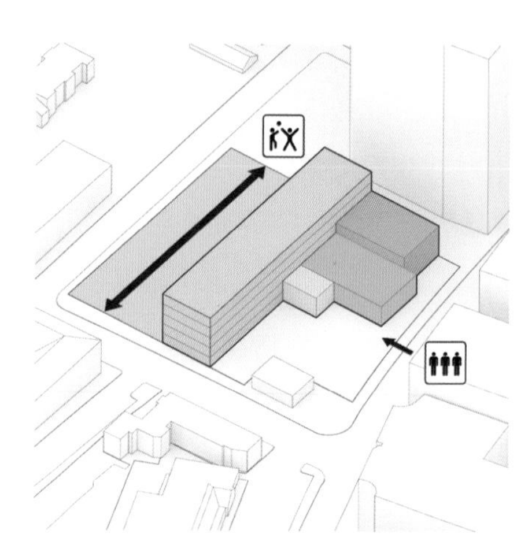

场地目标

SITE OBJECTIVE

A vertical stack of classroom bars stretches across the center of the site, creating a protective barrier between the athletic field and the busy urban corridor of Wilson Boulevard. Larger indoor spaces and community programs are placed on the Wilson and Quinn Street frontages.

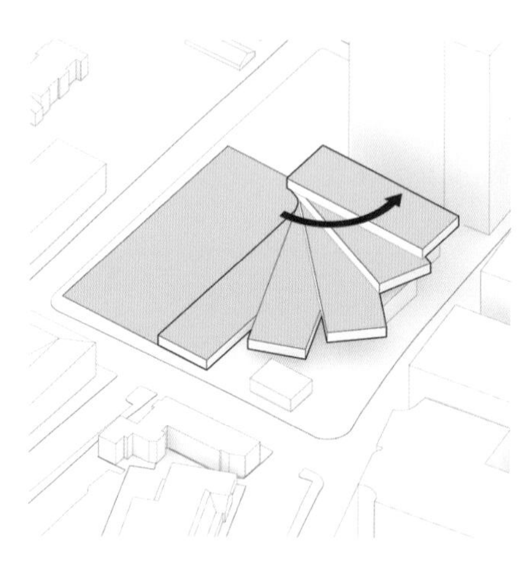

逐层降低的露台

CASCADING TERRACES

To create green space adjacent to the classrooms, the bars are rotated around a single hinge point. This creates 4 cascading terraces leading from the instructional spaces of the school to the recreation field. The first terrace will be accessible to the public when school is not in session.

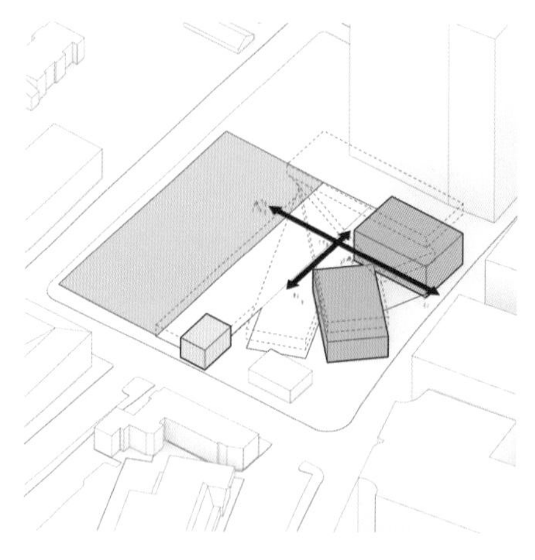

面向威尔逊大道的公共空间

PUBLIC SPACES TOWARDS WILSON

A generous lobby at the Wilson Boulevard entry invites the public inside for easy access to community-oriented programs hosted throughout the building. The gymnasium and auditorium are centrally located to the lobby.

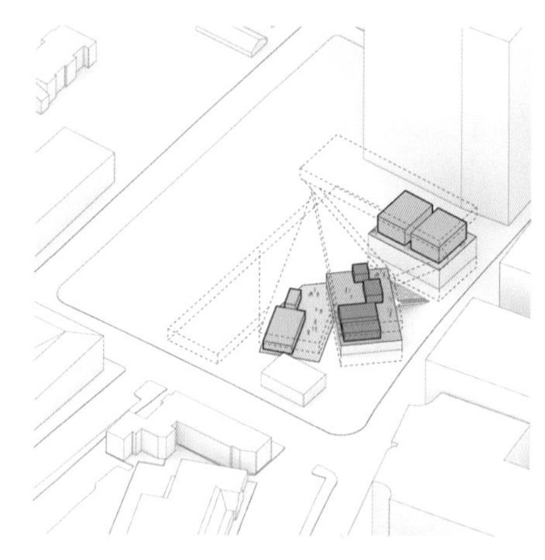

中间空间

IN-BETWEEN SPACES

Larger communal spaces such as the cafeteria, library, and music rehearsal rooms are easily accessible from the central space. The library is above the gym and the music rehearsal rooms are above the theater.

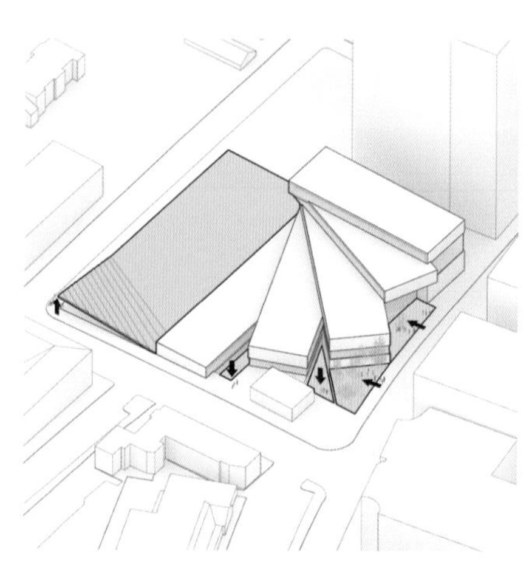

可操控一层平面

MANIPULATED GROUND PLANE

Slight manipulations to the ground plane create significant circulation and daylight benefits for the school. Two sunken courtyards provide quiet outdoor gathering spaces for supervised students while allowing natural light to illuminate the ground level. The outdoor spaces near the entrance to the school and another facing the corner of Wilson and Quinn Street serve as small public parks.

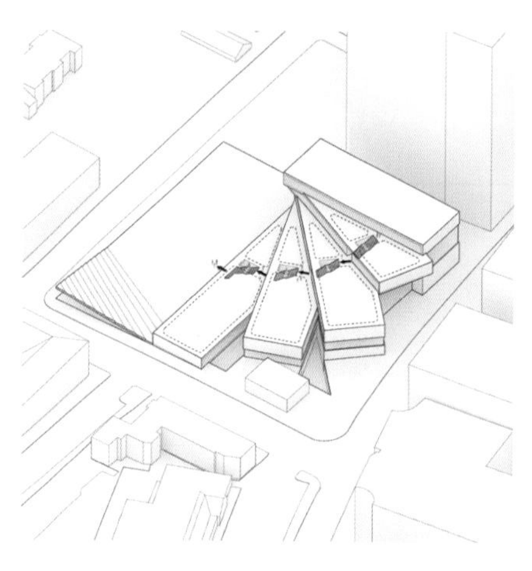

旋转楼梯连接

ROTATING STAIR CONNECTION

A rotating central staircase cuts through the interior of the building to connect the four-tiered terraces, allowing students to circulate outside and forging a stronger bond between the neighborhood and the school.

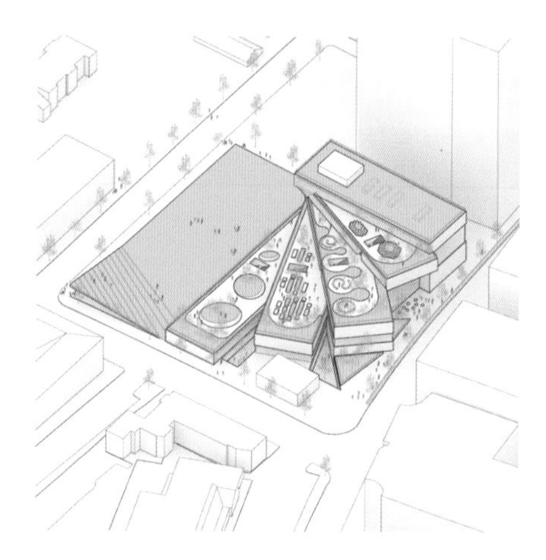

城市阶梯景观

URBAN TERRACED LANDSCAPE

Each of the 4 terraces provide different scales of activity, from large gatherings to class-size discussions and quiet study areas. These terraces give the opportunity for an urban school to have a 1-story feel that otherwise would not be possible.

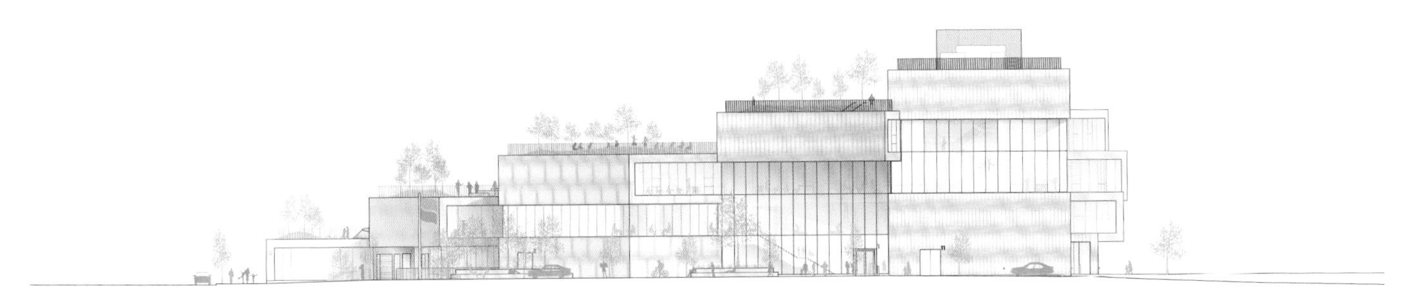

南立面 south elevation

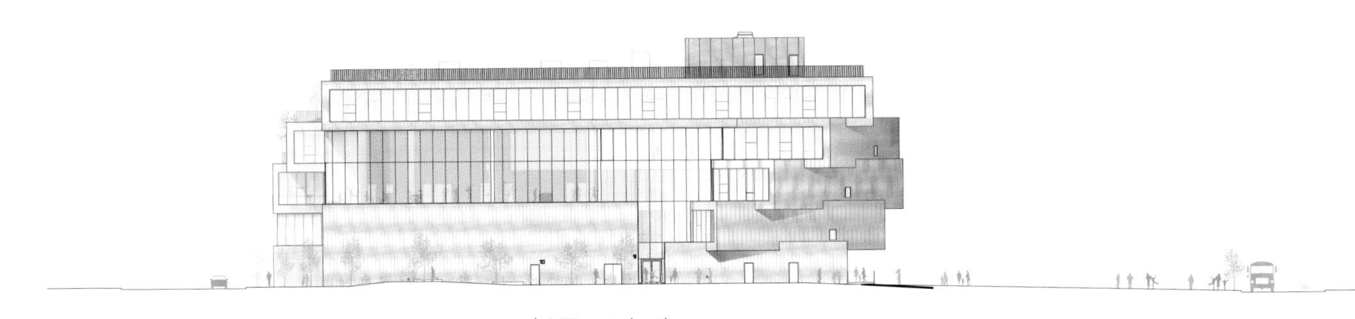

东立面 east elevation

北立面 north elevation

西立面 west elevation

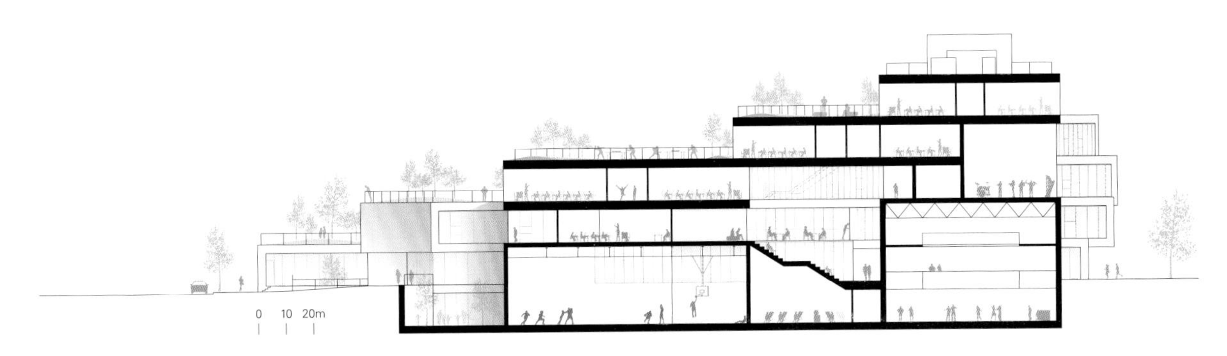

A-A' 剖面图 section A-A'

From Wilson Boulevard, students, teachers and staff are greeted by a triple-height lobby with stepped seating for use in assemblies and public gatherings. Many of the school's common spaces, including the 400-seat auditorium, gymnasium, library, reception and cafeteria, are directly adjacent to this lobby. Easy accessibility to the community-oriented programs hosted in the school encourages public interaction, creating a welcoming environment while heightening the visual connectivity between the shared spaces. Other specialized student spaces include an art studio, science and robotic labs, music rehearsal rooms and two performing arts theaters. The classroom bars serve as the primary organizing elements, surrounding a central vertical core that contains the elevators, stairs and bathrooms. As students enter from the central staircase, they are greeted by an expanded gradient of the color spectrum: each classroom bar is defined by its own color, combining intuitive wayfinding with a vibrant social atmosphere from the ground to the sky. The Shriver program, providing special education for students aged 11 to 22, occupies two floors of the building accessible from the ground floor. Specialized spaces dedicated to support APS' Functional Life Skills program are private and accessible: the gymnasium, courtyard, occupational physical therapy suite and sensory cottage are designed to aid in sensory processing. The Heights school's exterior is materialized in a graceful white glazed brick to unify the five volumes and highlight the oblique angles of the fanning classroom bars, allowing the sculptural form, the energy and the activity of the inside to take center stage. In keeping the surrounding neighborhood and former Wilson School in mind, the building's material palette pays homage to the historical architecture of Old Town Alexandria.

项目名称：The Heights School / 地点：Arlington, Virginia, USA / 事务所：BIG
合伙人负责人：Bjarke Ingels, Daniel Sundlin, Beat Schenk, Thomas Christoffersen
项目经理：Aran Coakley, Sean Franklin
项目负责人：Tony-Saba Shiber, Ji-young Yoon, Adam Sheraden
项目团队：Amina Blacksher, Anton Bashkaev, Benjamin Caldwell, Bennett Gale, Benson Chien, Cadence Bayley, Cristian Lera, Daisy Zhong, Deborah Campbell, Douglass Alligood, Elena Bresciani, Elnaz Rafati, Evan Rawn, Francesca Portesine, Ibrahim Salman, Jack Gamboa, Jan Leenknegt, Janice Rim, Jin Xin, Josiah Poland, Julie Kaufman, Kam Chi Cheng, Ku Hun Chung, Margherita Gistri, Maria Sole Bravo, Mark Rakhmanov, Mateusz Rek, Maureen Rahman, Nicholas Potts, Pablo Costa, Ricardo Palma, Robyne Some, Romea Muryn, Saecheol Oh, Seo Young Shin, Seth Byrum Shu Zhao, Sidonie Muller, Simon David, Tammy Teng, Terrence Chew, Valentina Mele, Vincenzo Polsinelli, Zach Walters, Ziad Shehab
合作者：LEO A DALY, Robert Silman Associates, Interface Engineering, Gordon, Theatre Projects, Jaffe Holden, Faithful+Gould, GHD, Hopkins Food Service, GeoConcepts, Haley Aldrich, The Sextant Group, Tillotson Design Associates, EHT Traceries, Lerch Bates, Sustainable Design Consulting
客户：Arlington Public Schools / 用途：education / 总楼面面积：16,700m²
竣工时间：2019 / 摄影师：©Laurian Ghinitoiu (courtesy of the architect)

1800 WILSON

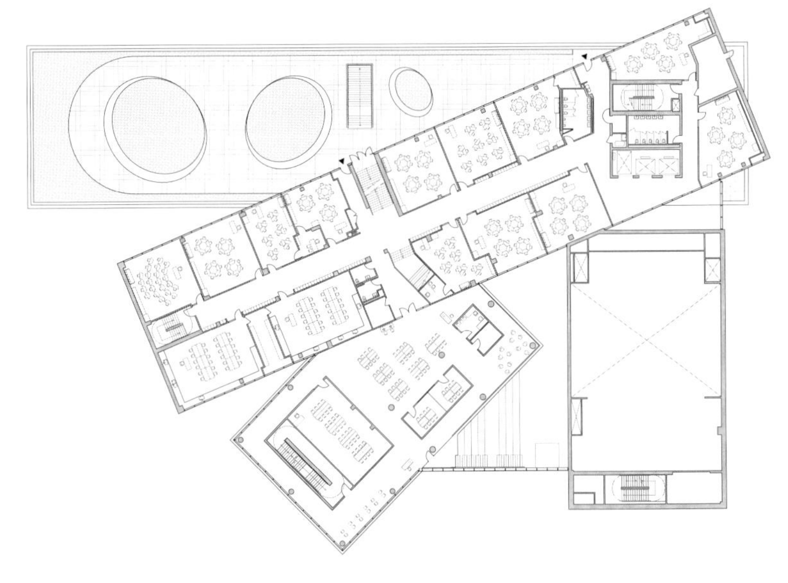

三层 second floor

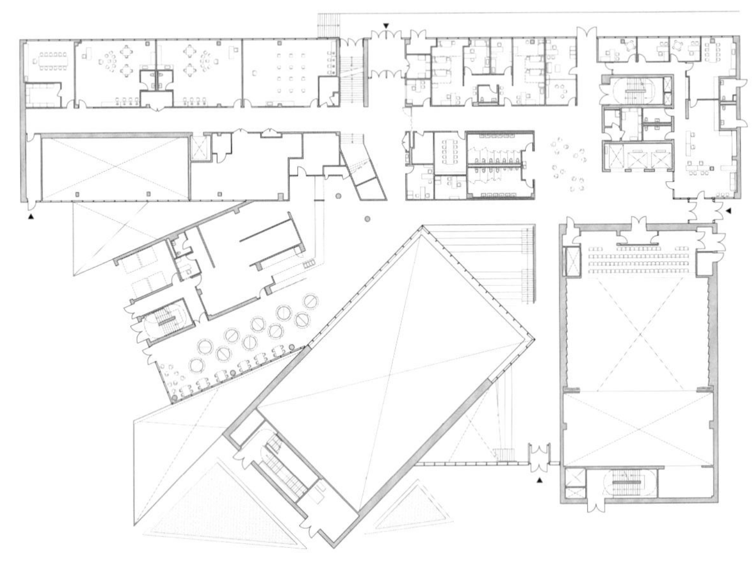

二层 first floor

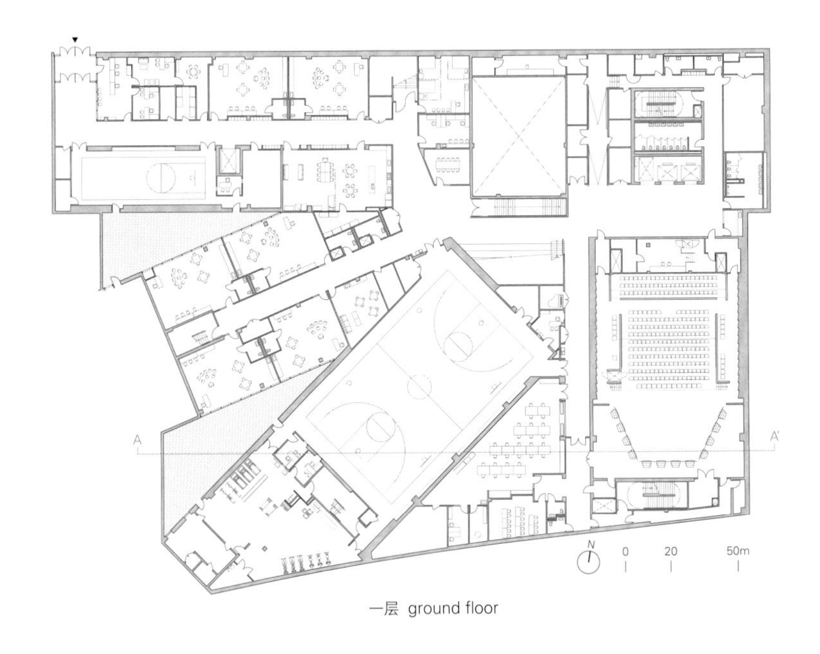

一层 ground floor

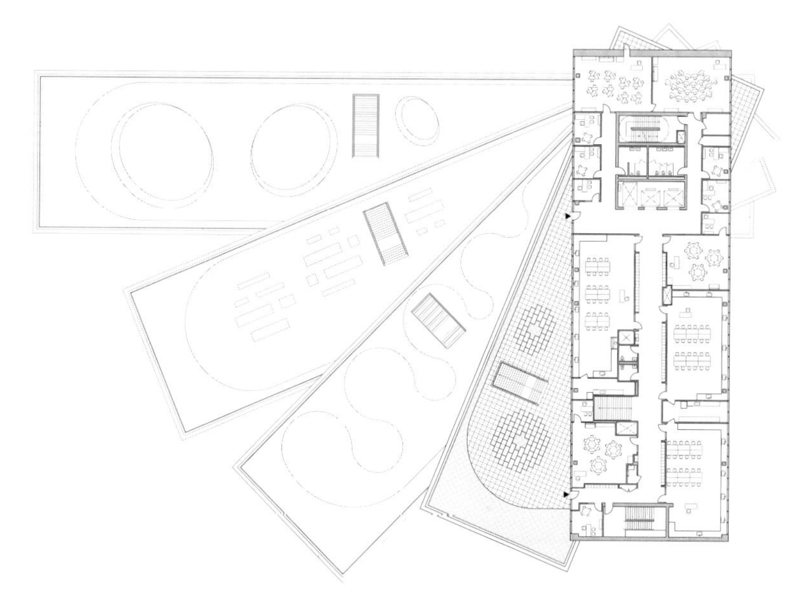

六层 fifth floor

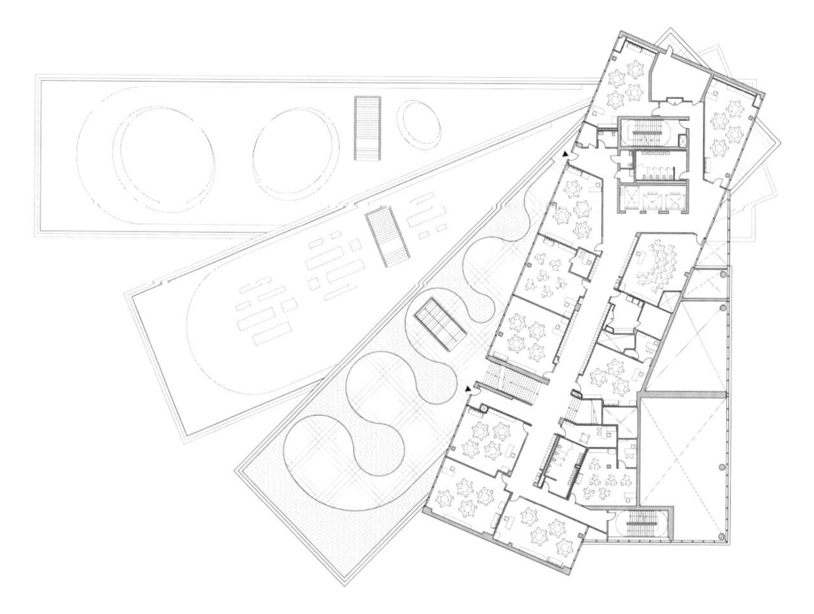

五层 fourth floor

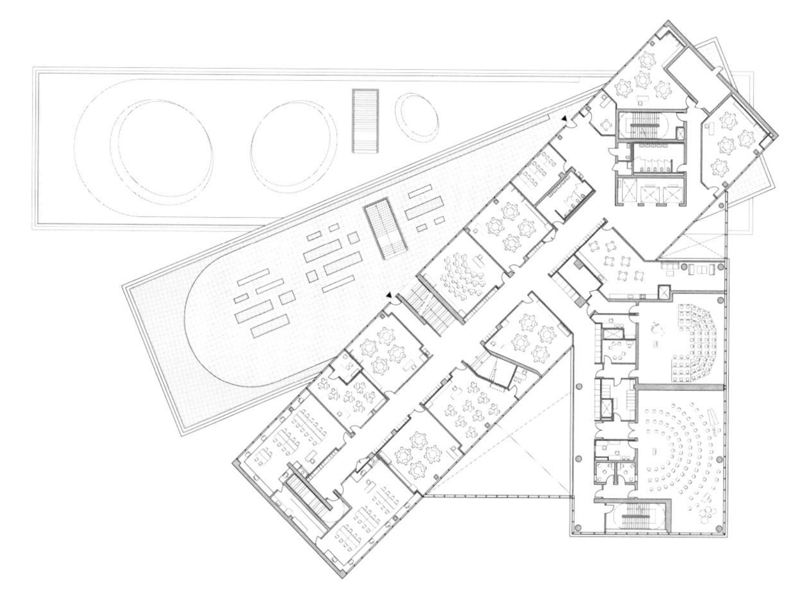

四层 third floor

办公场所

正如雷姆·库哈斯所言，“垃圾空间蕴含着完美未来的种子”。距离他在《OMA@work.a+u》上发表文章《垃圾空间》已经20年。从那时起，我们就开始关注工作空间的未来，以及人们对未来工作场所可能不断变化的期望。工作场所设计领域的重量级人物，时不时地宣称开放式办公空间的时代已经结束。但是真的会吗？开放空间具有的灵活性让人无法自拔。我们似乎无法摆脱开放空间所带来的灵活性。实际上，比起以前的任何时候，操作灵活性都更是我们现在的一部分。那么，在过去的20年里，我们对未来新工作场所的感觉和期望是否发生了很大的变化？确实有很大变化，但它又是如何变化的呢？

随着科学技术的进步，工作的性质必然发生变化，因为效率提高，所以人们有可能在更短的时间

As Rem Koolhaas proclaimed, “junkspace accommodates the seeds of future perfection.” It’s been 20 years since the publication of Junkspace, in *OMA @work.a+u*. We have since been preoccupied with the future of workspaces, and the evolving expectations of what the workplace of tomorrow might be. The corporate heavyweights in workplace design, every so often proclaim the end of the open office plan; but is it? We can’t seem to ween ourselves from flexibility that openness in space affords us. In fact, if anything, operational flexibility is more a part of who we are now than at any point in history. And so, have our perceptions, our expectations of the new workplace, in the place of tomorrow, changed all that much over the past two decades? Indeed, it has, but how?

The nature of work has certainly evolved with technological advancements, as efficiency has made it possible to complete far more tasks in shorter periods of time. Yet, the workplace is no

内完成更多的工作。然而，现在工作场所的重要性不亚于历史上任何时候。事实上，为了满足日常工作的基本需求，需要一个能够反映社交性的场所，这就要求更完整的内部体验。包罗万象的办公室就是这样的工作场所。

开放式办公空间的矛盾是真实存在的。人们渴望通透，希望沐浴在自然采光的开放空间中，或者在色彩绚丽的空间中与同事共事，这会充满平等的魅力。空间是完全开放的，但也被小心地保护着。工作场所看似布置简单，却是十分复杂。随着通信和交易的数字化，我们开始彻底地重新思量工作场所。在日常生活中，由活动边缘界定的物理空间的划分，以及一项活动和另一项活动的过渡，所有这些都是经过精心编排的。每一瞬间所使用的空间都被考虑在内。

less vital than at any point in history. In fact, the need for places reflecting our communicative societies, in satisfying our daily needs, argues for a more complete internalized experience. The all-inclusive office is the place to work.

The contradictions of open office space are real. We covet visibility, awash in bright streams of naturally daylit open or color saturated space amongst colleagues. with the appeal of equality. Spaces are open but are also carefully protected. The workplace is complex under the surface of seemingly simple arrangements. Our workplaces are being radically rethought as we operate in digitized worlds of communication and transaction. Divisions of physical space bound by active edges, transitional between one activity and the next, all carefully choreographed and complicit in our daily time. Every space wrapped in every moment is accounted for.

办公场所
To Work

Eric Reeder

（顾及其他场所的）办公

自工业革命以来，全世界的工作时间一直在持续减少。[1]技术改进、生产效率提高和更好的工作环境都大大缩短了工作时间。然而，当今对更高生产力和创造力的需求与我们所处的工作环境紧密相关。现在的工作空间不仅仅是作为工作空间存在的。最近由Balbek Bureau设计完成的Grammarly公司基辅办公室（108页）使人们对工作的另一面有了一些了解。半圆形的楼层规划，根据多种不同的功能分区，为普通的工作日提供动力。小憩室、隔声娱乐区、弯曲吊桥一侧的临时会议空间和休息空间，每个空间都为员工提供了"逃避工作"的场所。人们甚至可能会认为上班是为了休闲娱乐，这一点在Grammarly公司办公室的众多小面积功能区中得到了体现。办公室的每个角落都可以进行各种活动。工作与能提高效率的新式消遣活动处于适当的平衡。Grammarly的活动区已经融合在一个密集的微城市中。交通流线的设计是为了鼓励从一个区域到另一个区域的偶遇。各个功能区的边缘通过交通流线和建筑外围护结构连接在一起，建立了与外面世界的视觉联系。设计的核心意图似乎一直是让人感到仿佛置身于另一个世界中，让人体验到一种独特的空间效果。

（普通结构中的）创新工作

据说成功的工作场所"能将我们与他人联系起来[2]"。现代人与同事之间的关系比以往任何时候都更紧密。但是，不管是物理距离还是技术差距，不同程度的分离都会成为合作的障碍。CMR医疗设备公司总部大楼（122页）是由英国利物浦一家刚成立不久的公司——WMB工作室设计的。CMR是一家总部位于英国剑桥的新兴医疗设备集团，在过去的几年里，其规模迅速扩大。

新的公司中心总部旨在提供研究、测试和协作工作空间，以支持最近分散在剑桥市各地独立运营的工作组。为了减小各个工作组之间的距离，这家快速发展的公司希望建立一个集中的设施，让不同的工作组能在同一屋檐下开展各种活动。办公空间位于剑桥郊区乡村边缘的一个不起眼的"农业大棚"内。当地的规划法规要求新的工业建筑必须融入传统的农业景观

Working (with other places in mind)

Working hours around the world have been in steady decline since the industrial revolution.[1] Technology, productive efficiency, and better working environments have all contributed to reducing the amount of time workers spend on doing tasks. Yet, demands today for even better productive creativity are closely tied to the work environments we inhabit. Workspace now extends functionally beyond the expected tasks of work alone. The recently completed Grammarly Office in Kyiv (p.108) designed by Balbek Bureau offers a glimpse at the other side of work. The semi-circular floor plan is zoned with a diversity of programs to interrupt an average workday. Nap rooms, soundproof recreation areas, improvisational meeting spaces (parked along a meandering pedestrian bridge), and a lounge space, each providing places of escape. One might think we go to work for recreation, as demonstrated in the extensively granularized program of Grammerly. Each square meter of the office is planned for various engagement. Work is balanced with new forms of productive distraction.
Active zones at Grammarly have been fused in a dense micro-city. Circulation is intentionally designed to encourage improvisational encounters flowing from one area to the next. Programs are bound at the edges by the circulation and the building envelope, establishing view portals to worlds beyond. As if being somewhere else is always the central intent for the spaces with experiential effect.

Innovation Work (within a generic structure)

It has been suggested that successful workspaces "connect us with others[2]". We are more connected than ever with those that we work with. Still, degrees of separation, be they physical distance or technological gaps, become barriers for collaboration. The CMR Surgical HQ Building (p.122) was designed by the young design firm, WMB Studio in Liverpool, UK. CMR Surgical was founded as a start-up medical group based in Cambridge, UK and has expanded exponentially over the past several years.
The new central headquarter was designed to provide research, testing and collaborative workspaces supporting groups that have most recently operated in scattered facilities around the city of Cambridge. In seeking to bridge distance divides, the fast-growing company sought to establish a central facility for diverse range of activities under one roof. The space is housed within a non-descript "agricultural shed" at the rural

结构中，因此，设计参考了基本的农业建筑结构形式。

通用的钢框架壳体营造了宽敞和开放的内部空间，可供多人协作和个人独立工作使用。团队会议空间、个人休闲活动空间和制造实验室空间都沿着一条中央通道布置并连接在一起。设计师将阶梯式的中央中庭设计为通道顶点，这里是公共场所，阶梯式座位的安排增强了互动和交流。

设计师沿着中央通道创建了超饱和色彩的会议室，从而激活了整体的空间体验。休息区和活动休闲区通过玻璃墙在视觉上相互连接在一起。例如，攀岩墙和会议室就被并排设置，让员工被动参与进来。颜色和材料在单调和鲜艳之间形成鲜明的视觉对比，为空间增添了许多乐趣。

办公空间是（城市的）延伸

办公室是城市的延伸，或者用AMAA建筑师事务所（144页）的话说，“与城市、城市的历史和城市的层次产生深厚联系”。我们知道，寻找过去和现在众多层次之间的联系是建筑师的责任。通过想象彼此之间（在建筑中）的因果关系，可以将这些不同的层次合并为一个。我们解析历史，试图寻找多种方式将物理形式、材料安排、结构改进和机械系统联系起来，以满足现代的舒适性要求，同时融合、分离和突显昨天和今天的区别。

AMAA建筑师事务所在意大利阿尔齐尼亚诺设计的工作室在很多方面都表达了对双重时间的敬意。新的办公空间悬挂在旧厂房内。老旧的工厂代表着另一个时代的工业产品，混凝土和砌体结构外加钢支撑代表的是当今的成果；两个时代彼此成就。新的玻璃箱式办公区是一个透明的空间，位于旧厂房的中间，所有人都能看到。新旧材料都裸露在外，并且都未加修饰，显示出材料本来的样子。两个截然不同的时代同时呈现在人们眼前。建筑师们小心地布置工作空间、交通流线空间和支持设施的每个构件，使它们彼此之间产生独特的联系。所有这一切都是可见的。

城市的未来和城市工作的中心都取决于是否愿意超越无形空间的划分。我们需要认识到空间动态同时性下的办公表现

urban fringe of Cambridge. Local planning regulation required that new industrial structures fit into the fabric of a traditionally agricultural landscape, hence the basic forms of agrarian reference.

A generic steel-framed shell provides generous interior open space to test collaborative and individual choice. Spaces for team meetings, individual leisure activity, and fabrication labs are stitched together along a central circulation pathway. The circulatory road, as designed by the architects, culminates in a terraced central atrium, known as the common place for interaction and congenial connection enhanced by the stepped seating arrangement.

The designers created hyper-saturated color meeting rooms along the central pathway activating the overall spatial experience. Combinations of sedate program and active recreation areas are visually connected through glass walls. A climbing wall and a conference room for example, are positioned side by side in passive participation. Colors and materials create enhanced moments with striking visual contrast between mundane and bright surfaces.

Work extensions (of the city)

The office is an extension of the city or in the words of the architects AMAA (p.144), "weaving a deep connection with the city, its history and its layers". We know as architects this is a responsible act, seeking connections between a multitude of layers portraying the past and present. Such layers can be integrated as we imagine reciprocal cause and effect (in architectural terms). We unpack history and find ways to connect physical forms, material arrangements, structural enhancements, mechanical systems for modern-day comforts, merging, isolating and highlighting yesterday and today.

The design of Studio in Arzignano, Italy by AMAA, in many ways is homage to a duplicity in time. New office space is suspended in old factory. The aging factory represents industrial productions and work from another era. A concrete and masonry structure is supplemented with added steel support for today's performance; each epoch casts one against the other. A new glass box office area is situated in the middle of the old factory, a transparent space for all to see. New and old materials are left exposed and unrefined in natural states. Two distinct times read side by side. The architects are careful to place each element of working area, circulation and

多样性。可以说，我们在设计中需要纳入对未来的考虑，让其成为不常见的混合动力。其他空间中具有替代功能的空间建立了以前不存在的新关系，在密封的旧建筑中创建出前所未有且充满活力的空间新形式。到目前为止，仅通过材料的组合就实现了这一切。

（随处可进行的）适应性工作

工作文化正在改变。经济变化改变了人们的工作方式和工作地点，因此，适合各类型协作工作的场所重建了新经济时代下的空间需求。19世纪伦敦的狗岛曾是造船等行业的中心，但工业搬迁导致它被废弃。该区域最近再次发展，成为住宅和办公场所理想的再开发场地。但是现在所面临的挑战（和机遇），就是随着旧工业的搬迁，老旧建筑中留下了许多空间。如今，适应性再利用是对旧工业区合理且负责任的处理方式。

艾默里斯建筑师事务所为一家名为Craft Central的非营利慈善机构设计了The Forge办公和展览空间（132页）。Craft Central为富有创造力的专业手工艺人提供他们负担得起的工作室空间。该项目是在一座经过翻新的19世纪工厂厂房内，建造设备齐全的"独立插件"，这些插入结构被布置成堆叠的两层工作室，沿着厂房的长轴方向设置。

线性布置的"吧台"插入结构被细分为小型工作室，两端是展览空间和杂物空间。工作室悬在工厂结构内，工厂地板上方的空间提供了展览、通道以及陈列空间，这些空间布置在工作室的周围。线性的插入结构表面装饰裸露的胶合板，这有利于营造过去与现在之间的材料协调感。

材料的重要性

研究表明，生产力与工作场所设计密切相关。一些分析师推测，在具有更好设计的空间里，认知任务的效率可以提高25%。[3]其中，空间布置与材料的选择密切相关。材料会影响日常工作的体验和感受。

utilitarian support in distinct relation to each other, all within visual reach.
The future of cities and our urban centers of work will rely on willingness to transcend implied division. We need to recognize multiplicities of performance in dynamic simultaneity. The future will need to be embedded, so to speak, mixed as uncommon hybrids. Spaces of alternative functions within other spaces, establish new and unfounded relationships – the suspension of dynamic new forms within the hermetic old. For now, this takes form almost exclusively as material constructs.

Adaptable work (anywhere)

The culture of work is changing. Shifting economies have evolved how and where we work. Thus, places suited to different types of cooperative service have reestablished spatial needs in new economic times. The Isle of Dogs in London was once home to shipbuilders amongst other 19th century industries. Industrial relocation led to abandonment, today having made a dramatic comeback as desirable redevelopment location for live and work. Notwithstanding the challenges (as well as opportunities) as old industries relocate leaving behind vacuums of space in aging buildings. Adaptive reuse is now the logical and responsible action for aging industrial areas.
The Forge Offices and Exhibition Space, designed by Emrys Architects (p.132), was planned for the non-profit charity organization known as Craft Central. The organization supports affordable studio spaces for creative and specialized craft professionals. Self-containing "stand-alone insertions" operate as stacked, 2-story studios situated lengthwise within the renovated 19th century factory shell.
The linear "bar" inserts are subdivided into small working studios and bookended with exhibition and utility spaces. Studios float within the industrial shell and void spaces of the factory floor provide exhibition, circulation and display areas around the perimeter of the working studios. The linear inserts are finished in exposed plywood panel contributing to a sense of material coordination between past and present.

Materials Work

Studies suggest that productivity is tied to workplace design. Cognitive task productivity may increase by 25% in better designed spaces, some analysists speculate.[3] Part of this equation, spatial arrangement is connected with material choice. Materials shape experiential and sensorial aspects of the working day.
Second Home London Fields (p.92), located in East London's Hackney working-class neighborhood, was designed with materiality – inside and outside – in mind. A unifying facade membrane of tautly stretched ETFE,

伦敦菲尔兹Second Home项目（92页）位于东伦敦的哈克尼工人阶级社区，在设计时就特别考虑了建筑内部和外部的材料。统一的外立面膜由绷紧拉伸的ETFE材料制成，在多样的功能空间中提供了多种内部设计可能性。设计灵感来自弗雷·奥托的外表皮呈乳白色半透明状，从街道上可以看到，外表皮悬浮在半空中，带有精致的钢框架。随机放置的圆形门板点缀着紧绷的表面。其半透明的特质让光线充斥着整个内部共享的协同工作空间。

这座建筑是一个改造后的社区剧场，现在的演出内容取决于不同背景的年轻创意工作者们。郁郁葱葱的绿植墙壁隔开了由基本的隔声材料包裹的空间。建筑中有一个新的舞台装置，能够吸收并有选择地传播日常活动的声音。柔和的有机绿色和光滑的彩色硬质表面之间的对比，可以很好地弱化不同区域之间的差别。Second Home项目的设计保持了视觉上的开放和工作区之间的互通，看上去并不会妨碍充满活力的使用者。

工作空间未来蓝图

在今天工作就是在未来工作。我们的空间和具有不确定性的时间需要一个有适应性的框架。随着技术的发展，我们的认知也在变化，我们认识到环境变化带来了挑战，也出现了很多新的工作方式，在这种演变中，空间将一直需要灵活的支持。我们可以随时随地与任何人一起工作。我们不需要与工作伙伴面对面。数字世界可以将我们带到需要去的地方。那么，这对未来的物理空间、我们的工作场所以及我们的城市有什么影响呢？

也许未来的工作会需要更少的空间界定。现在我们可以或已经开始反思对过去的工业建筑中固定的静态场所进行掏空和拆除处理的做法是否正确。我们看到现在许多西方国家城市的传统店面和工厂被掏空，因此更有必要重新定义临时使用这一概念了。回想起来，"垃圾空间"是真实存在的。这是过去的残余，因为我们不断地对其进行废弃再利用。在此，我们有机会重新设想一下如何管理城市结构，以及最终如何管理工作场所。现在，我们有了一个新机会，那就是在工作中调整各种不同的组合。

holds together a vast array of interior possibility in diversified programming. The Frei Otto inspired exterior skin, visible from the street, undulates of milky white transparency, suspended in mid-air with a delicate steel frame. Randomly placed circular portals punctuate the taut surface. Its transparent quality allows light to flood the shared co-working spaces inside.

The building is a transformed community theater, whose performances now rest on a variety of young creative workers from different backgrounds. Walls of lush plants divide spaces wrapped in basic, acoustically minded materials. A new stage set is capable of absorbing and selectively transmitting the sounds of daily activities. The contrast between soft organic greens and smooth colored, hardened surfaces is just enough to damper differences from area to area. Second Home remains visually open and accessible from workspace to workspace, seemingly without compromise to its dynamic inhabitants.

@ Work

To be at work now is to be at work in the future. Our spaces and our uncertain times are requiring a framework for adaptive spirit. As we evolve with technology and recognize the challenges of environmental change, with new means of work, spaces in this evolution will continue needing the support of flexible prospect. We can work anywhere in the world at any time, and with anyone. We need not meet face to face with those we work with. The world of the digital takes us to where we need to be. What does this hold for the future of physical space, our places of work and in essence our cities?

To be at work tomorrow may even require less spatial definition. We can and have begun to rethink the hollowing out and deconstructing of fixed and static places of past industries. As we witness the emptying of traditional storefronts and factories now in many western cities, redefining temporary occupation may become even more necessary. In retrospect, Junkspace is real. It is the residual of the past as we are constantly faced with repurposing the discarded. In this there is opportunity to reimagine how we manage the fabric of our cities and ultimately our places of work. We now have a new opportunity for shaping far different combinations at work.

1. Max Roser, "Working Hours", *OurWorldInData*, 2020, Retrieved from: https://ourworldindata.org/working-hours
2. "The Future of Workplace", *Wired Insider*, Nov. 2018, https://www.wired.com/brandlab/2018/11/the-future-of-workspace/
3. Shama Hyder, "Research Says Companies That Do This 1 Thing Increase Worker Productivity by 25 Percent", *INC*, Oct. 12 2017, Retrieved from: https://www.inc.com/shama-hyder/research-shows-that-companies-that-do-this-one-thing-increase-worker-productivity-by-25.html

伦敦菲尔兹Second Home项目
Second Home London Fields

Cano Lasso Architects

METROPOLITAN DRINKING FOUNTAIN & CATTLE TROUGH ASSOCIATION

半透明外立面在伦敦菲尔兹最新潮的创意共享办公空间中体现出感知的作用

A translucent facade plays with perceptions in London Fields' trendiest new creative co-working space

伦敦菲尔兹以前是伦敦东部哈克尼市的工人阶级社区，由于年轻的中产阶级设计师和艺术家们负担不起市中心昂贵的租金，都搬到了这里，因此现在这里已经成为一个受他们欢迎的地区。

Second Home项目入驻伦敦菲尔兹，凭借其多元化的合作计划，免费提供多功能空间、咖啡厅、托儿所等服务，促进了当地创意社区的兴起。

该建筑原本是19世纪的一个舞厅，对其进行的改造主要集中在两个基本部分：外立面和室内。在外立面上，必须提供一个新的参照点，将形象从传统转换为新潮；而在室内，应用"幻觉理论"，帮助摆脱单调乏味，创造一个不同的室内空间。

Second Home多元化的方案和创新精神造就了无数可能，尤其是将老剧场有限的原始构件进行升级后，更加让人眼前一亮。

外立面覆盖材料设计经历了不同的变化过程。最初参考了塞德里克·普莱斯在20世纪60年代为伦敦动物园鸟舍所做的设计，该鸟舍是一个简单的金属网结构，由内部的轻型结构支撑，既若隐若现，呈现透明状，又可以阻隔目光从几乎隐形的金属网中透过。但这一想法被驳回了，因为鸟舍覆盖材料显得微妙的根本是鸟舍拥有宽敞的空间。

因此，建筑师们选择了ETFE膜，其半透明质感与其遮挡的内部实体建筑更加和谐。为了让这种膜成型，设计师们采用了从弗雷·奥托的实验过程中学到的一种方法——直觉优先，这种方法使设计师可以根据随机放置的刚性点来生成建筑的形状。

与Second Home的设计理念一样，建筑中所有事物的设计都有幻想、发明和制造的元素：轻盈和脆弱性、透明度、配色方案以及对自然的借鉴。通过使用自然构造学，建筑师们运用了大自然的逻辑和欧氏几何逻辑。

建筑师写道："我们对针对惯性、重力和刚性空间的自然构造学十分感兴趣；建筑是……感知的本质。在伦敦菲尔兹，我们尝试用可塑性的表达方式来探索周围景观，从感性和经验来看，这种可塑性似乎会产生其他空间和现象，可能是自然的，但是却拥有另一种规模。如此，空间就会变成一种氛围体验和景观体验。"

开放式的空间还可以通过划分区域来单独使用。不同的子空间通过家具、植物和单元空间来界定，它们都是通透式的隔断，可以灵活地改变空间的大小，只需要简单地挪动一下位置，就能创造出多种可能的布局方式。

光线也因不同的区域得到了增强，因此，人们对建筑的感知也是在不断变化的，不仅仅是在路线上，在时间上也是如此。这会让建筑中的使用者感觉身处居住景观之中，同时还会发现感知的诗意美。

London Fields is an ex-working-class neighborhood of Hackney, east London, which has become a popular and alternative district for young middle-class designers and artists who can't afford the expensive rentals of the inner city.

Second Home's arrival to London Fields, with its diversified co-working program and free provision of multipurpose spaces, a cafe and nursery, contributes to the emergence of the creative local community in the area.

Originally a 19th century dance hall, the building's adjustment to the program focused on two basic interventions: first, the facade, which had to offer a new reference point and a change of image from the traditional to the trendy; and second, a different interior applying the "illusion theory", offering an escape from boredom.

Second Home's diversified program and innovative spirit allow for numerous possibilities, especially when combined with the enhancement of the limited original elements of the old theater.

东立面 east elevation

北立面 north elevation

西立面 west elevation

南立面 south elevation

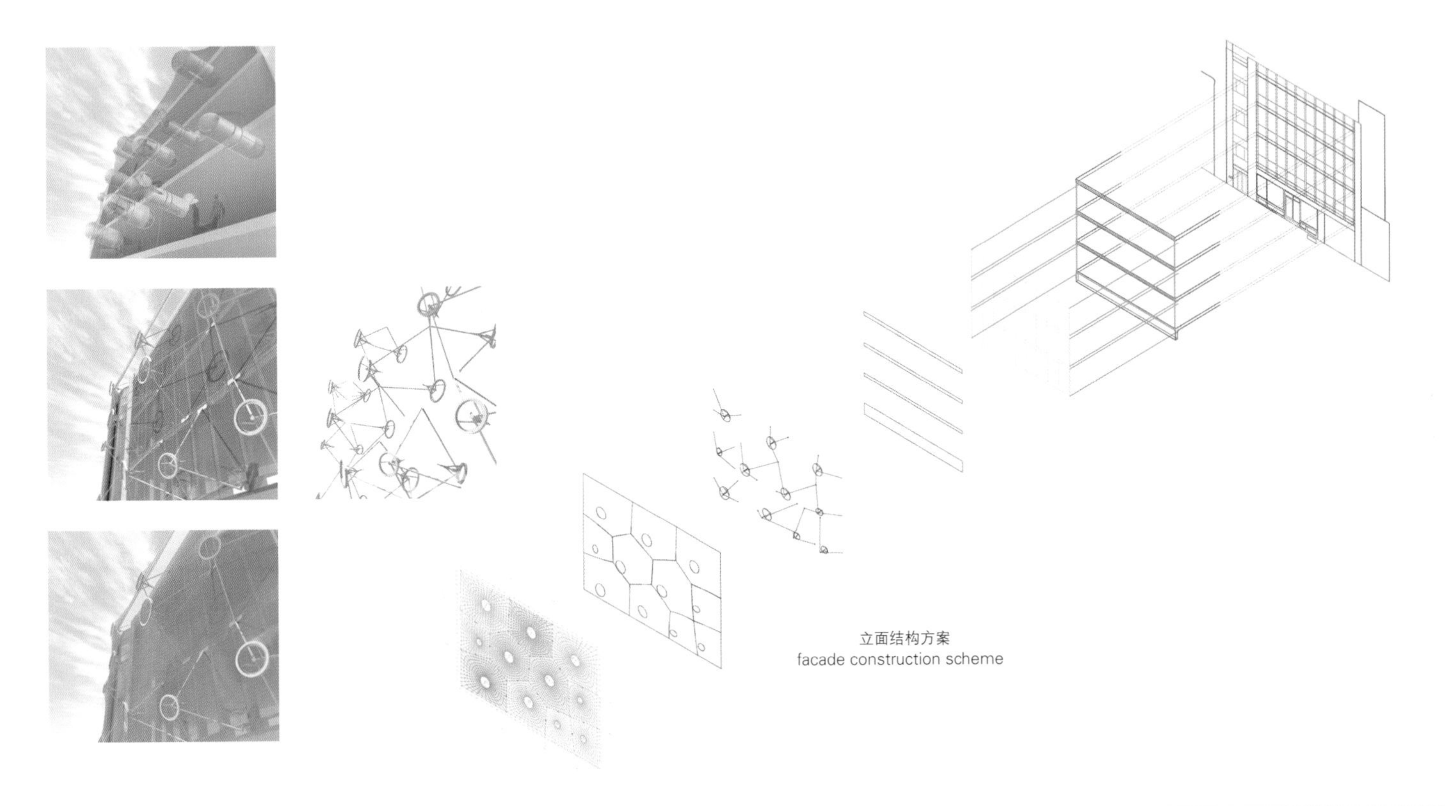

立面结构方案
facade construction scheme

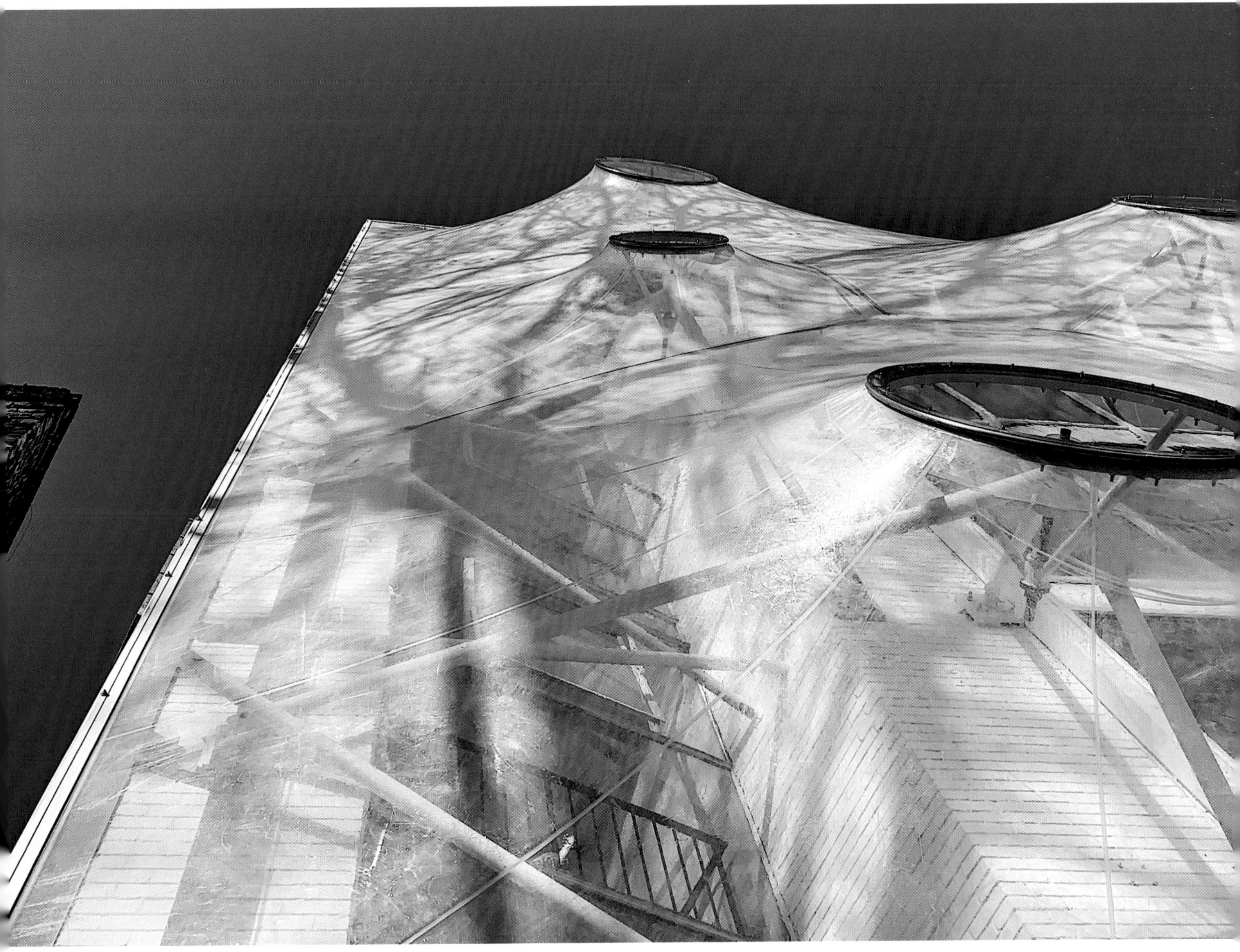

The facade's veiling went through a process of different materialities. The original reference was Cedric Price's 1960s aviary for London Zoo, a simple metallic mesh stretched by a lightweight structure based on internal supports, which created an intriguing and subtle transparency that filtered the gaze through the almost invisible mesh. But this idea was dismissed as the airy void of the aviary was intrinsic to its veil's subtlety.
Consequently, the architects opted for an ETFE membrane; its translucency works better with the solid volume it hides. To give shape to this membrane, an approach learnt from Frei Otto's experimental processes – intuition – allowed the designers to generate a topography based on rigid points placed randomly.
Consistent with the spirit of Second Home, everything has an element of fantasy, invention and manufacture: the lightness and fragility, the transparency, the color scheme, and the references to nature. Working with natural tectonics allowed the architects to engage with the logic of nature and the synthetic logic of geometry.
"We are interested in nature's tectonics facing inertia, gravity and rigid spaces; architecture is... the essence of perception. In London Fields, we've tried to explore the landscape with a plastic expression which, from our sensibility and experience, appears to generate other spaces and phenomena, maybe natural, with another scale. This way, the space becomes an atmospheric experience and a landscape experience", the architects write.
The open plan spaces also allow for individually disparate uses through the provision of delimited zones. Different sub-spaces are defined by furniture, plants and cells created by see-through partitions. These elements allow for spatial contractions and expansions is with a simple gesture, endless possibilities of arrangement arise.
Light too is intensified by areas, so that the perception of the building is constantly changing, not only route-wise but also time-wise. This awakens in the building's users an awareness of being in the middle of a living landscape and a discovery of the poetry of perception.

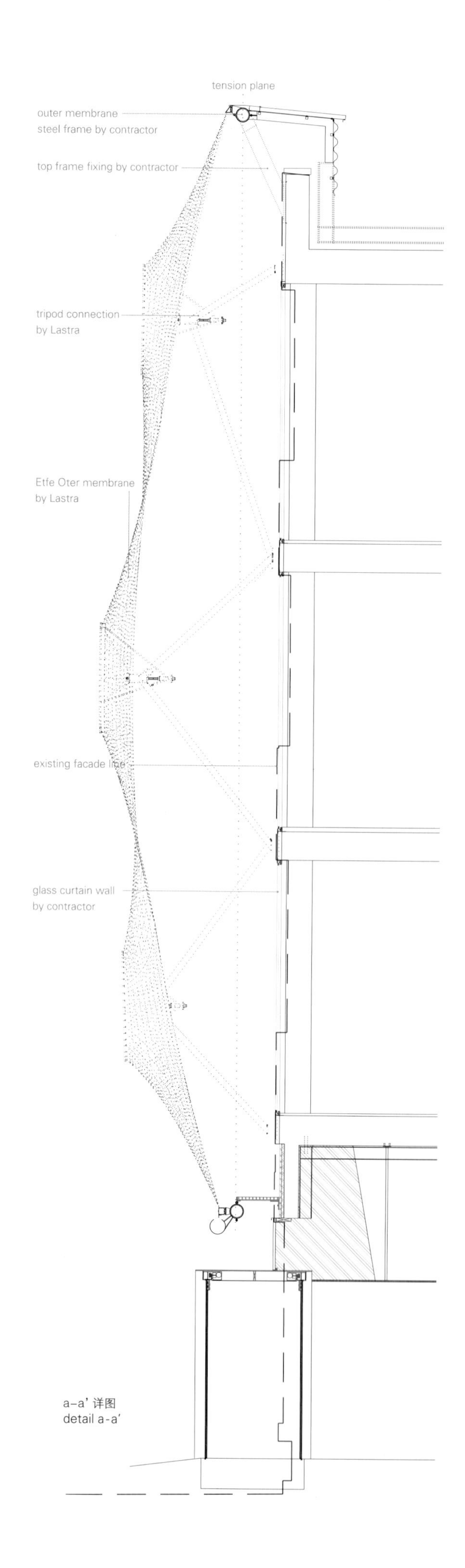

a–a' 详图
detail a-a'

项目名称：Second Home London Fields
地点：Mare Street 125-127, Hackney, London, UK
建筑师：Cano Lasso
建筑师负责人：Gonzalo Cano Pintos
设计团队：Gonzalo Cano Pintos, Diego Cano Pintos, Alfonso Cano Pintos
客户：Second Home
工程师：Webb Yates Engineers / 景观设计：Noel Kingsbury
顾问：EXA Group UK Ltd. / 合作者：Alfonso Nebot, Ignacio de la Vega Copado, Rosa Cano Cortés, Gerardo Martín, Ana Pardo, Ana Olalquiaga Cubillo, Rocío Marina Pemán, Carlota Galán Daries, Alfonso Cano Abarca
建筑面积：1,812.98m² / 竣工时间：2019
摄影师：©Iwan Baan (courtesy of the architect) - p.92~82, p.98, p.97 top, p.102, p.103 top-left, p.106~107; ©Second Home (courtesy of the architect) - p.95, p.97, p.99 bottom-left, bottom-right, p.103 top-right, bottom, p.105

Platforms POWERED ACCESS
HA260PX

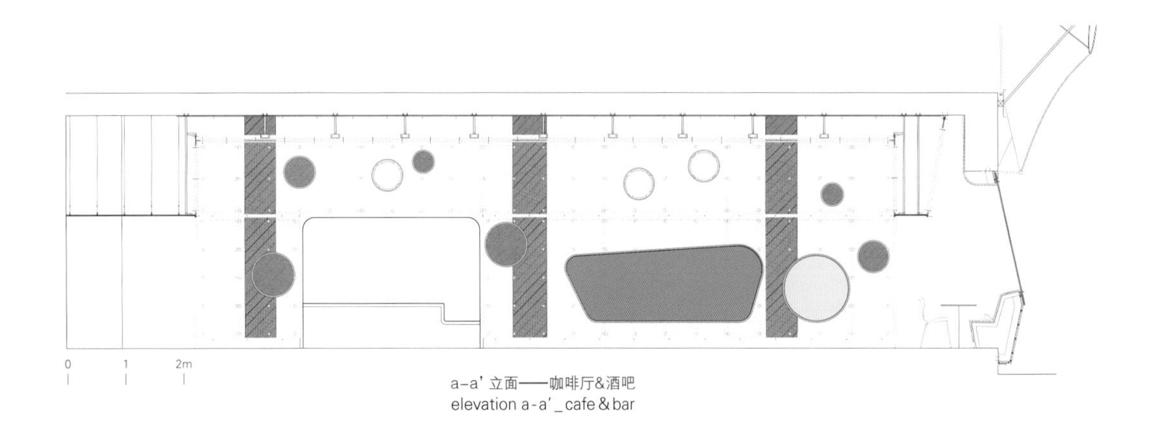

a–a' 立面——咖啡厅&酒吧
elevation a-a' _cafe & bar

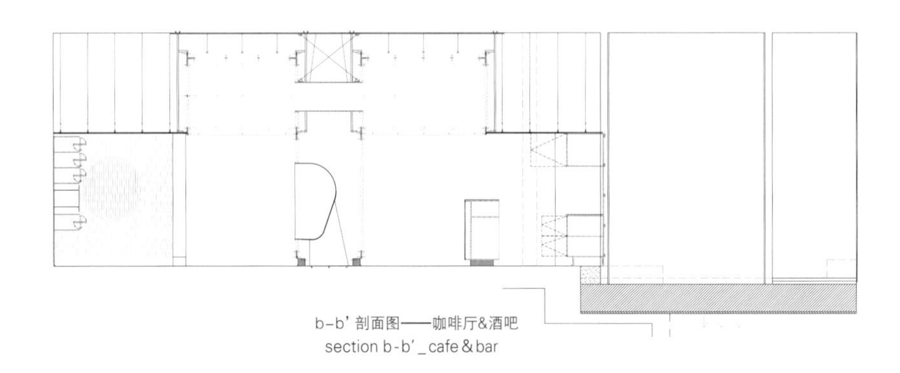

b–b' 剖面图——咖啡厅&酒吧
section b-b' _cafe & bar

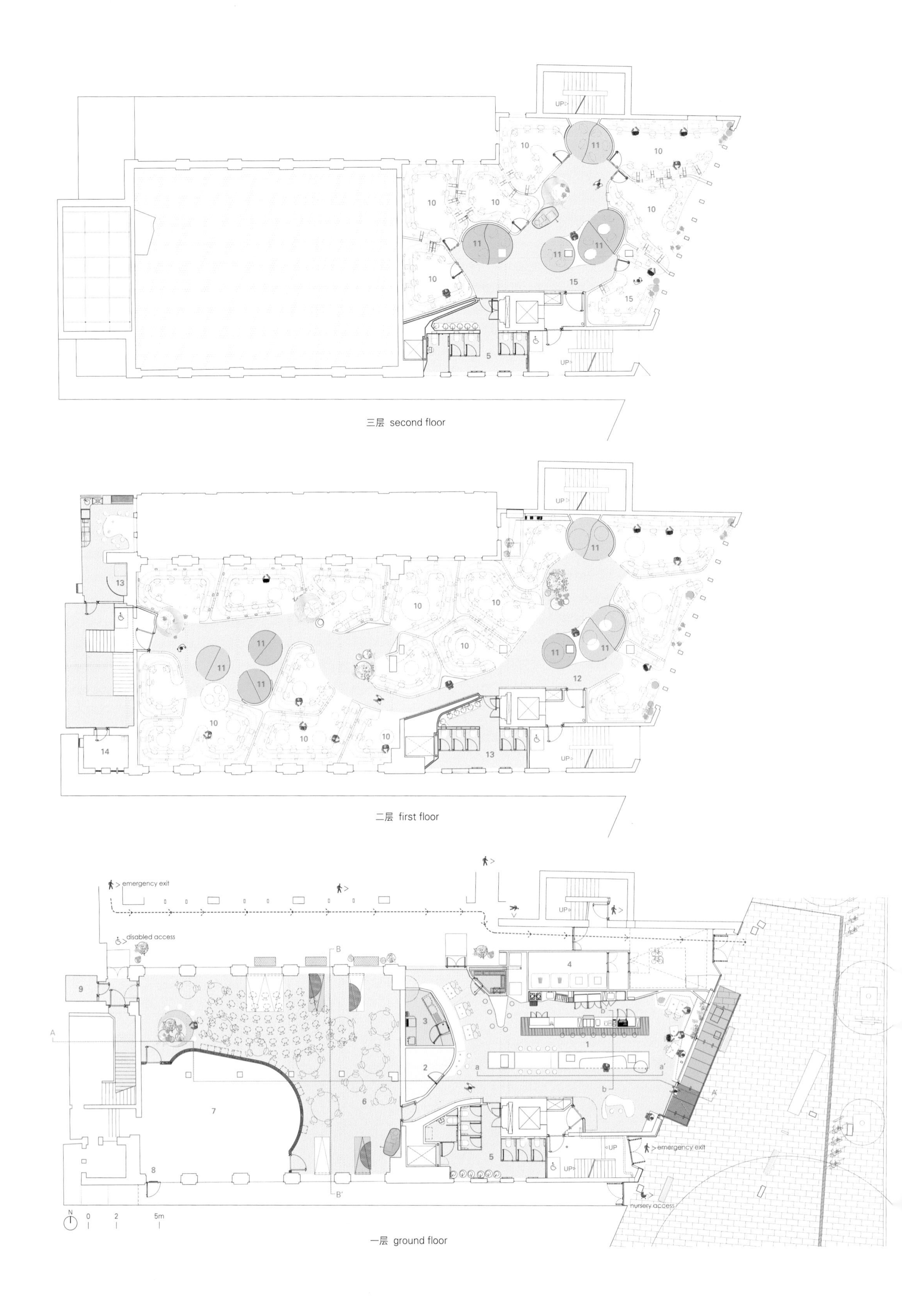

三层 second floor

二层 first floor

一层 ground floor

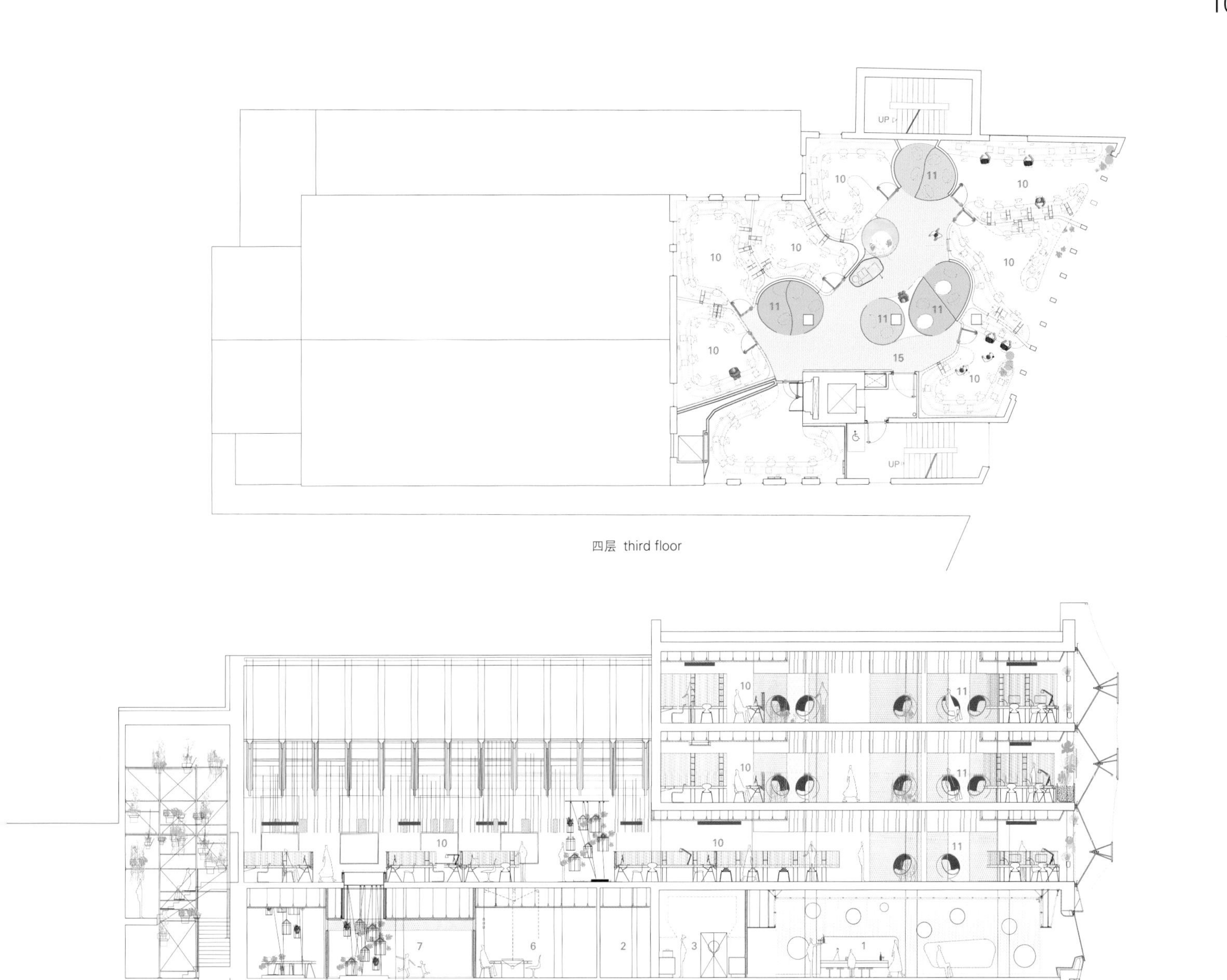

四层 third floor

A–A' 剖面图 section A-A'

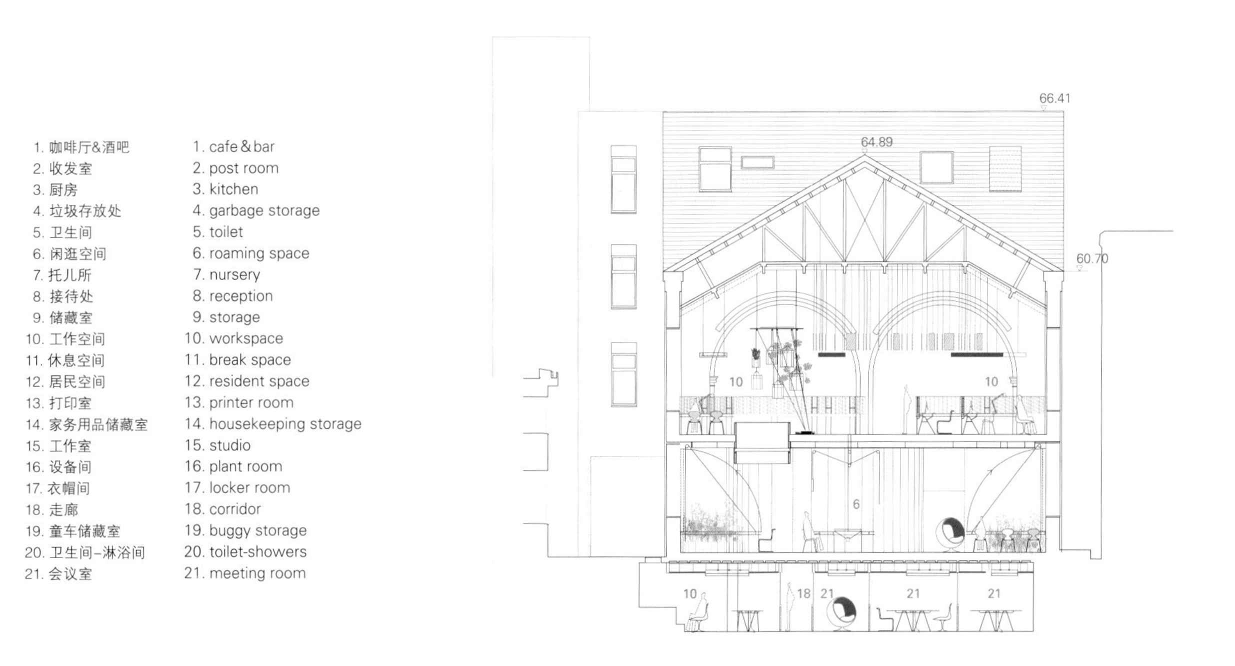

B–B' 剖面图 section B-B'

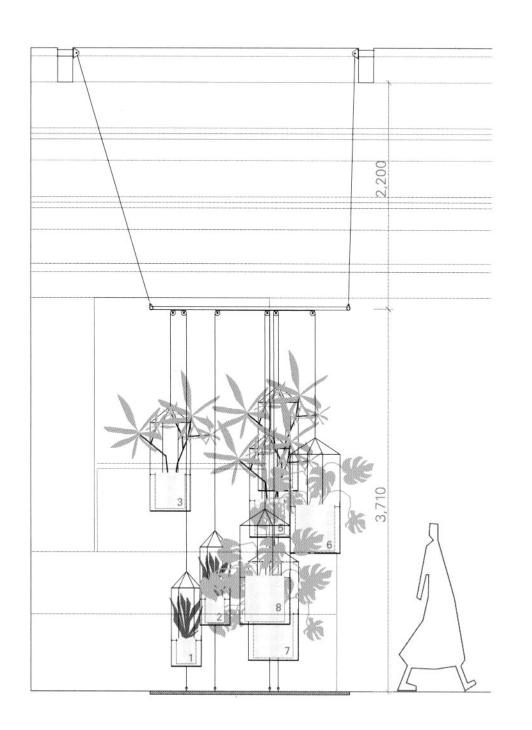

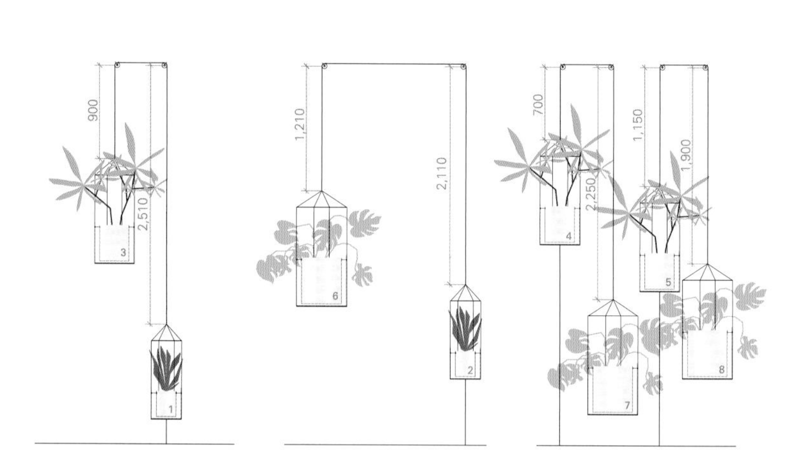

悬挂系统 hanging system

30 mm methacrylate panel. Screwed on plasterboard structure

Fixing partition formed with Knauf 52/25 U chanels, with 50/35 C studs @ 400 mm centres, clad each side with 2 layers of 15 mm Knauf Performance Plus Panel. 50 mm rockwool insulation inside. Inner layer continuing to floor below.

Plaster finishing on both side. All sides finished in white paint. Colour (BS) 4800 00-E-55

Artigo rubber flooring. MULTIFLOOR ND UNI - Colour U105 on both sides.

Existing slab and structure. Cut in between beams to form void.
New floor build-up by structural engineer

12.5 mm plasterboard suspended ceiling. Finished with a 25 mm layer of dark natural cork at 2.20 from FFL

Double fixing partition formed with Knauf 52/25 U chanels, with 50/35 C studs @ 400 mm centres, clad each side with 2 layers of 15 mm Knauf Performance Plus Panel. 50 mm rockwool insulation in creche's side.
(15+15+52+15+15+52+15+15)

12.5 mm plasterboard suspended ceiling at 2.60 from FFL. Finished with 40 mm white melamine acoustic wedged foam

2 x 20 mm methacrylate screwed to 40x40x4 L profiles, fixed with allen screws. 30 mm fixations on both ends of each panel. Steel fixed to plasterboard substructure
5 mm steel cylinder embedded in wall at various heights, fixed to platerboard reinforcement substructure.
Painted in various colours:.
Colour 01. (BS) 381 593
Colour 02. (BS) 0-008
Colour 03. (BS) 4800 00-E-55

Hidraulic spring to open lid

Swinging 15 mm plywood disk finished with polished 0.8 mm steel sheet

Plaster finishing on both side. Roaming space's finished in white paint. Colour (BS) 4800 00-E-55.
Creche's side finished in white melamine acoustice wedged foam

Warm linear LED light in plasterboard niche

Artigo rubber flooring. MULTIFLOOR ND UNI - Color U108. Curved against inner plasterboard layer to form skirting, with profile PS25

Artigo rubber flooring. MULTIFLOOR ND UNI - Color U105. Curved against inner plasterboard layer to form skirting, with profile PS25

Grammarly公司基辅办公室
Grammarly Office in Kyiv

Balbek Bureau

90m弯曲吊桥创造出方便沟通与交流的办公空间
A workspace for communication and exchange created by the 90-meter bridge in a gentle curve

Grammarly是一家提供数字化写作辅助并帮助用户实现清晰有效沟通的跨国公司，在旧金山、纽约和基辅都设有办事处。该公司全天候运营，广泛使用IT设备，并运用了大量的通信和数据交换，无论在个人工作环境还是团队工作环境中都发挥着巨大作用。

2016年，Grammarly基辅分公司团队不断壮大，并决定搬到更大的空间中。公司的运营性质决定公司需要为不同类型的活动提供大量空间，包括：一个可容纳150个座位的大型会议厅，用于演讲和演示（又名聚会区）；许多配备IT设备的小型会议室，用于在全球各地的办公室之间进行高质量音频和视频会议；接待区；隔音娱乐室；员工食堂；小憩室；几个休息区以及设施区。

其他要求包括：使用生态友好型材料，室内采用暖色调，营造家的氛围，采取高科技会议连接方式和电气化系统，空间具有灵活性和适应性。

原有的办公区在一栋14层办公楼的顶层，现在分为两层：一层的天花板高8.8m，开放空间面积为$1300m^2$；夹层天花板高3m，面积为$450m^2$。

设计师的布局思路是将会客区放在一层，使其成为办公室的核心元素，并在会客区的周围设置六个具有隔声特性的开放式工作区。

夹层的布局经过调整后，拥有正确的曲率，能够将工作区和会客区分隔开来。这两层通过新安装的开放式楼梯相连。

几个独立的盒式房间是为了保护隐私而设的，可以作为安静的工作区，也适用于一对一交流。较大的会议室可容纳8至15人，分布在会客区周围，且每间会议室颜色不同。

办公室内有21个休息区，包括图书馆、小憩室和可以玩乐器或看视频的音响室，甚至还有增加舒适度和放松度的壁炉区，这里也可以作为非正式的聚会场所。

办公室大量采用生态友好型材料和天然木材，比如，办公空间中创造性地再利用了橡木材料，通过多种方式参考了休息区的户外空间，充分使用自然光，这些都有助于为员工创造工作与生活的积极平衡，并打造一个高舒适度和高效率的工作环境。

小憩室包含三个睡眠空间。从概念上讲，这三个是由窗帘封闭的空间，窗帘还可以遮挡光线。这三个空间装有床垫，床垫下有重量传感

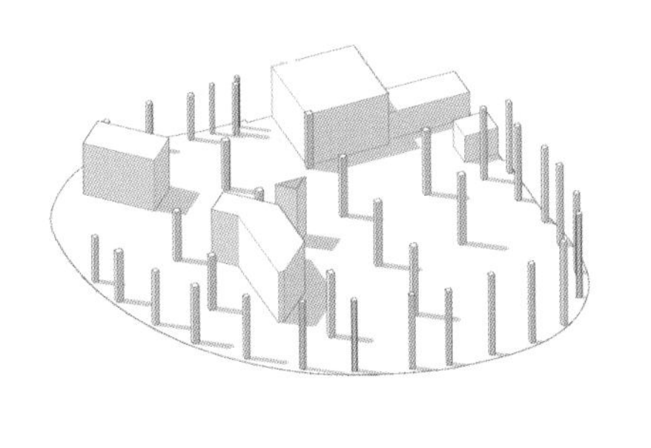

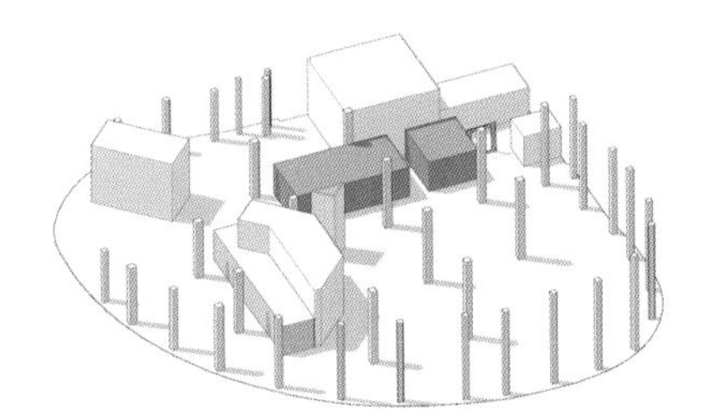

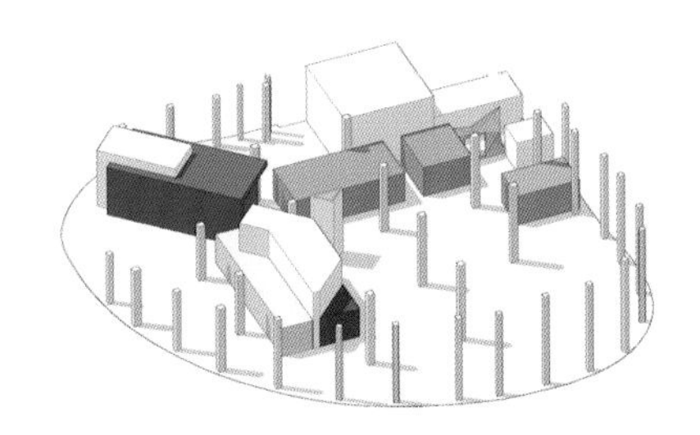

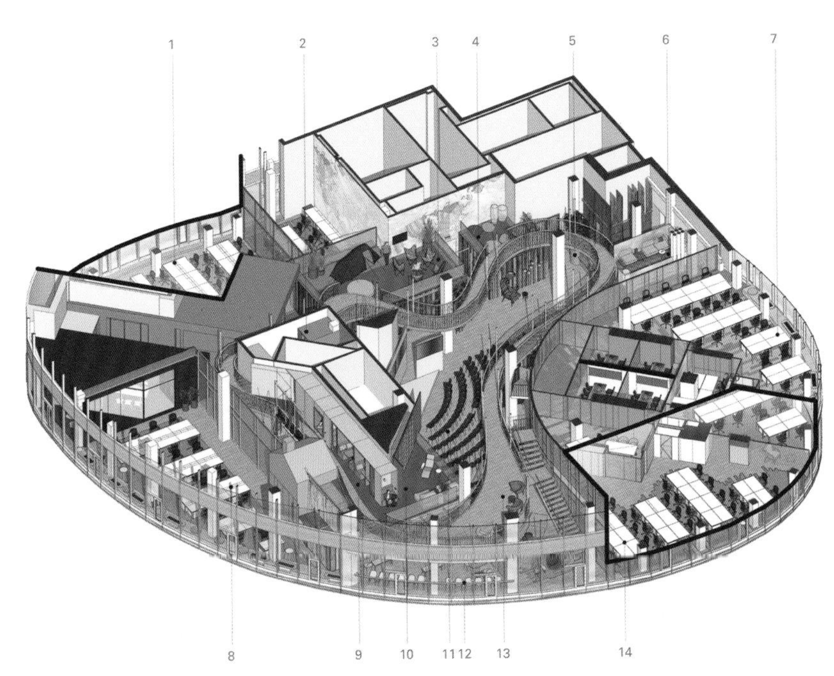

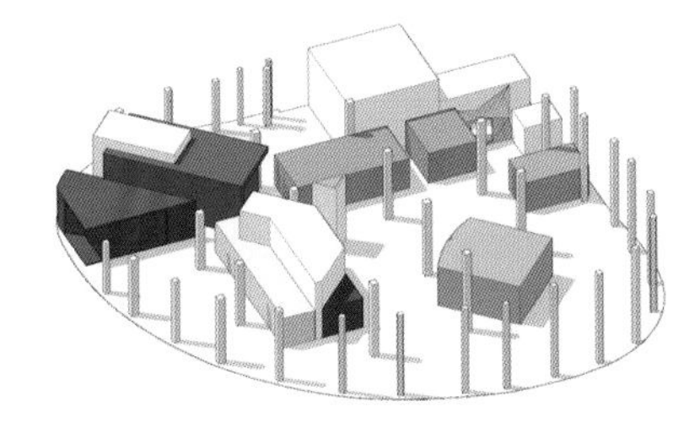

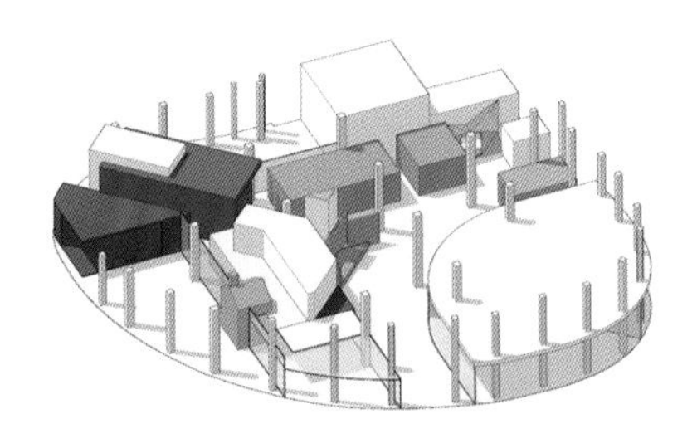

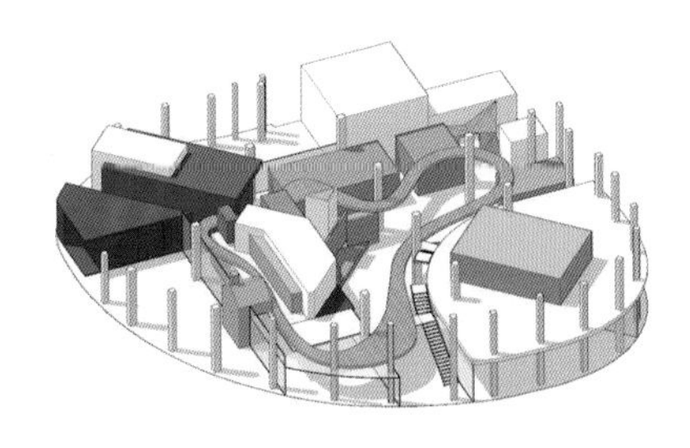

1. 开放式办公室，蓝色 2. 小憩室 3. 壁炉区 4. 石材休息区 5. 接待处 6. 木艺休息区 7. 二层开放式办公室，酸橙色 8. 开放式办公室，绿色 9. "Platzkart"休息区 10. 图书馆 11. 会客区 12. 厨房 13. 吊桥 14. 一层开放式办公室，沙色

1. open-plan office, blue 2. nap room 3. fireplace lounge 4. lounge with stones 5. reception 6. woodcraft lounge 7. first floor open plan office, lime 8. open plan office, green 9. "Platzkart" lounge 10. library 11. meet-up 12. kitchen 13. bridge 14. ground floor open plan office, sand

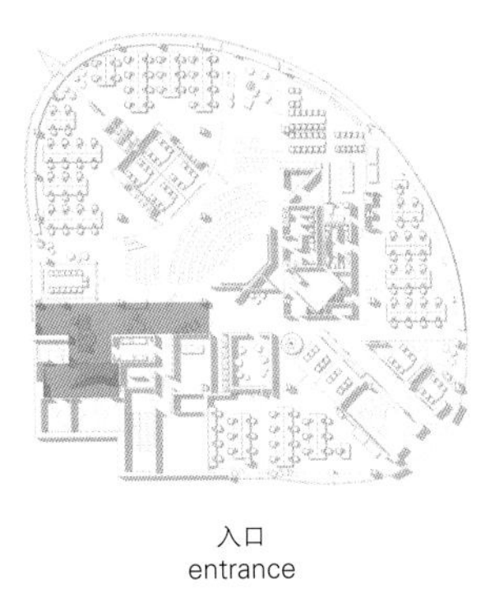

入口
entrance

设置黄色会议室的办公空间
workspace with
the yellow meeting rooms

设置深绿色会议室的办公空间
workspace with the
dark green meeting rooms

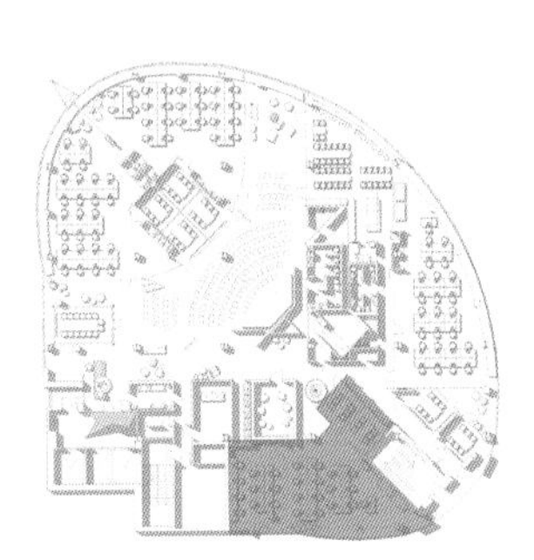

办公空间
workspace

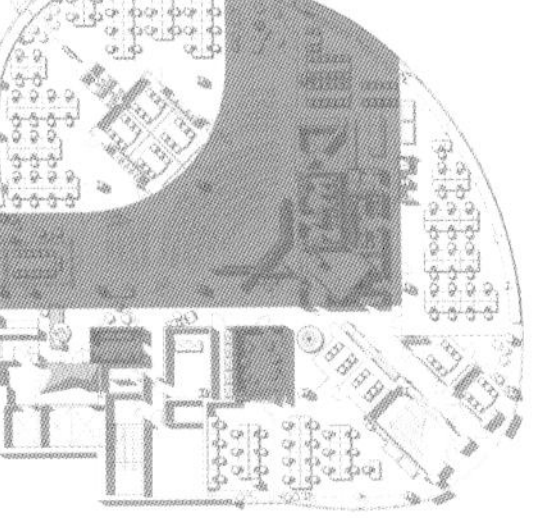

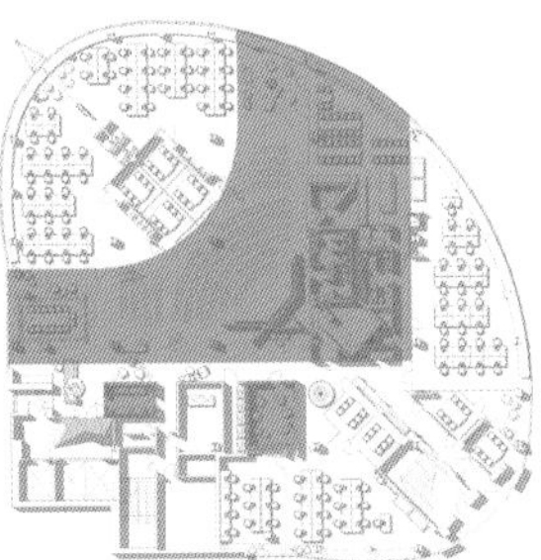

会客区&公共区
meet-up & public

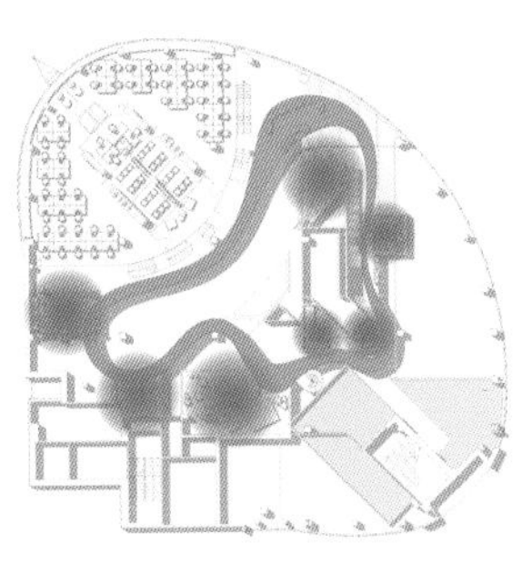

吊桥
the bridge

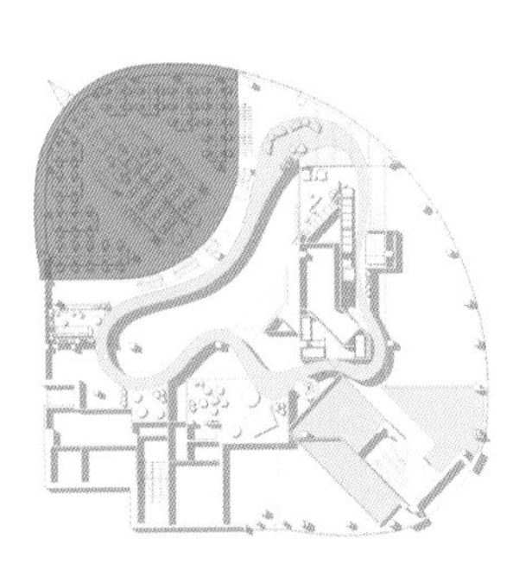

设置酸橙色会议室的工作空间
workspace with the lime meeting rooms

器。如果有人在睡觉，传感器就会做出反应，对灯光指示器发出"Zzz"信号，表示该区域处于使用状态。

一层房间上方的休息区通过办公室的关键元素之一——吊桥，与工作区相连。90m长的吊桥以平缓的曲线环绕办公室，俯瞰整个办公室，并稍稍向外扩展，将它周围的休息区也收纳进来。吊桥悬挂在天花板上，地面没有支撑。为了将桥梁的厚度降到最低，设计师将洒水系统管道设置在主楼板下方，分管道与吊桥结构构件混在一起融入桥体中。建造办公室用了50t金属，其中25t用于建造吊桥。吊桥将办公室的各个功能区统一起来，使这些功能区拥有相同的生命周期或节奏；吊桥也象征着位于基辅、纽约和旧金山各分公司之间不可分割的关系。

Grammarly – a company providing a digital writing assistant to assist users in making their communication clear and effective – is a global company with offices in San Francisco, New York, and Kyiv. The company operates 24/7, with the extensive use of IT devices and a high volume of communication and data exchange, both within individual and group settings.

In 2016, the Grammarly Kyiv team grew and decided to move to a bigger space. Due to the nature of operations, the company required a variety of spaces for different activity types, including: a large conference hall with 150 seating capacity, for lectures and presentations (aka a meet-up zone); a number of small meeting rooms equipped with IT devices for quality audio and video conferences between offices across the globe; a reception zone; a soundproof recreation room; a canteen for employees; a nap room and several lounge zones; and facilities areas.

Other requirements included eco-friendly materials, a warm color palette and homely feel for the interior, high-tech solutions for conference connection and electrification systems, as well as flexibility and adaptability of space.

The original office area, occupying the top floor of a 14-story office building, now consists of two levels, with the ground level featuring an 8.8m high ceiling and 1,300m² of open space, and the mezzanine level with a 3m high ceiling and 450m² of space.

1. 开放空间，酸橙色
2. 休息区
3. 办公室
4. 会议室
5. 木艺休息区
6. 石材休息区
7. 壁炉区
8. 小憩室
9. 榻榻米休息区
10. “Platzkart”休息区
11. 图书馆
12. 吊桥

1. open space, lime
2. lounge
3. office
4. meeting rooms
5. woodcraft lounge
6. lounge with stones
7. fireplace lounge
8. nap room
9. Tatami lounge
10. “Platzkart” lounge
11. library
12. bridge

二层 first floor

1. 入口
2. 休息区
3. 开放空间，沙色
4. 可转变用途会议室
5. 会客区
6. 木艺休息区
7. 接待休息区
8. 衣柜区
9. 杂物间
10. 储藏室
11. 音响室
12. 开放空间，蓝色
13. 会议室
14. 开放空间，绿色
15. skype聊天室
16. 厨房
17. 打印区
18. 卫生间

1. entrance
2. lounge
3. open space, sand
4. transform meeting rooms
5. meet-up
6. woodcraft meeting room
7. reception wait zone
8. wardrobe
9. utility
10. storage
11. loud room
12. open space, blue
13. meeting rooms
14. open space, green
15. skype-booth
16. kitchen
17. print zone
18. WC

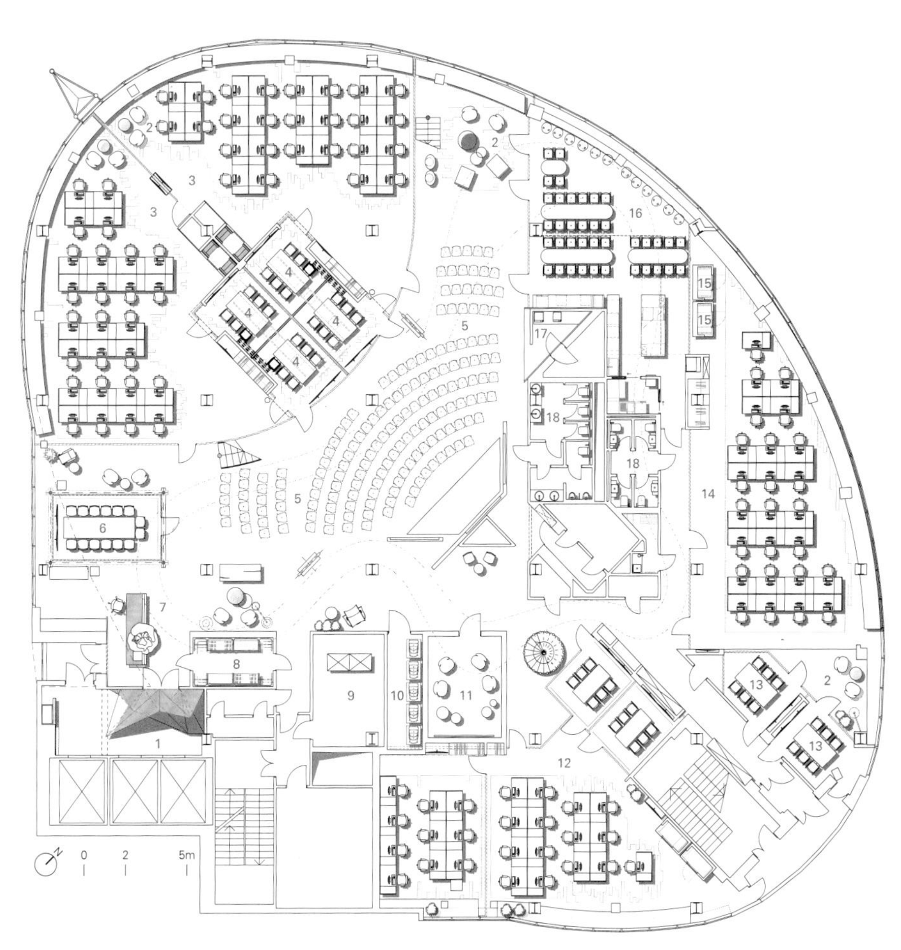

一层 ground floor

The designers' layout idea was to place the meet-up zone on the ground floor, making it a core element of the office around which six open space working zones, each with soundproof qualities, are located.
The configuration of the mezzanine floor was adapted to ensure the correct radial curve that separates the working areas from the meet-up zone. The two levels were connected by the newly installed open staircases.
Several individual box rooms offer privacy for quiet work and one-on-one communication. Larger meeting rooms, with capacities of 8 to 15 people, are located around the meet-up zone and are finished in different colors.

There are 21 lounge zones in the office, including a library, the nap room and an acoustic room to play musical instruments or watch videos. There is even a fireplace zone for added comfort and relaxation, which is also used as an informal meeting place.
Extensive use of eco-friendly materials and natural wood, including creative reuse of oak throughout the office space, multiple references to the outdoors in the lounge areas, and the abundance of natural light, all help to create a positive work-life balance for the employees and contribute to highest levels of comfort and efficiency.
The nap room accommodates three sleeping places. Concep-

tually, these are three blocks closed by the curtains for optical isolation. The blocks have mattresses under which weighing sensors are located. If someone occupies a sleeping place, the sensor responds and sends a "Zzz" signal to the light indicator, which means that the block is engaged.

Lounge areas created above the rooms of the ground floor are connected to the working zones by one of the key elements of the office – the suspension bridge. The 90m bridge loops around the office in a gentle curve, overlooking the entire office and expanding slightly to accommodate lounge zones in its path. The bridge has no ground support, it is suspended from the ceiling. In order to keep the thickness of the bridge to a minimum, sprinkler system pipes were passed under the main floor, and their fragments incorporated into the body of the bridge, blending with the structural elements.

Fifty tons of metal was used in the construction of the office, out of which 25 tons went into making the suspension bridge. It unites various functional zones of the office into a single life cycle or rhythm and has also become a powerful symbol of the inseparable connection between Grammarly offices in Kyiv, New York, and San Francisco.

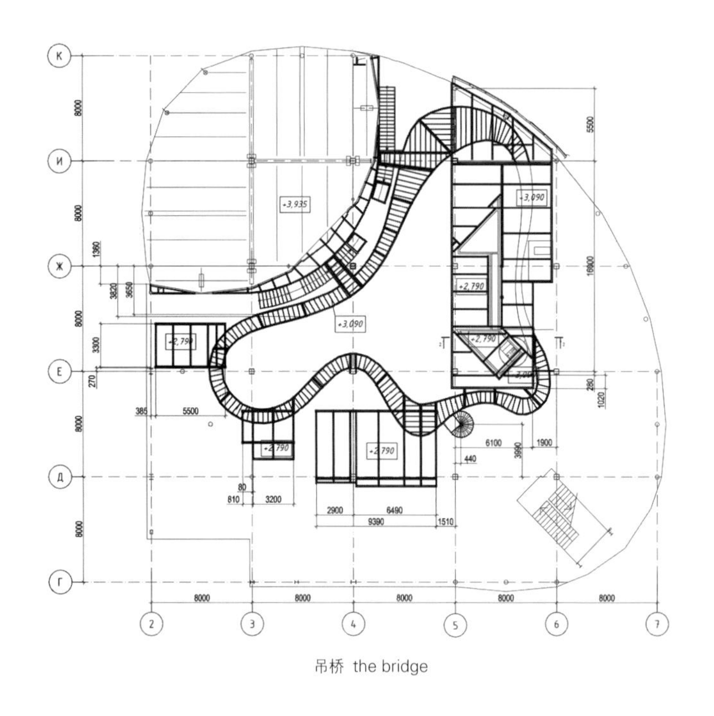

吊桥 the bridge

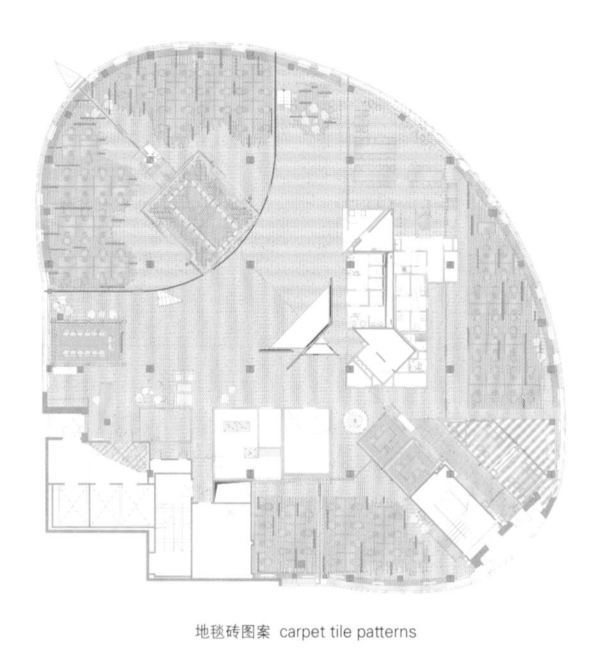
地毯砖图案 carpet tile patterns

项目名称：Grammarly Office in Kyiv / 地点：Kyiv, Ukraine / 事务所：balbek bureau – Slava Balbek, Andrii Berezynskyi, Anastasia Marchenko
项目经理：Borys Dorogov / 客户：Grammarly Kyiv / 设计时间：2016 / 竣工时间：2018.11
摄影师：©Andrey Bezuglov, Yevhenii Avramenko (courtesy of the architect)

Z
Z
Z

CMR医疗设备公司总部大楼
CMR Surgical HQ Building

WMB studio

由废弃仓库建筑改造的CMR医疗设备公司新总部成为高科技的摇篮

CMR Surgical's new HQ becomes a cradle of high technology in an abandoned shed building

WMB工作室已经完成了位于剑桥的CMR医疗设备公司总部大楼项目，CMR是一家领先的医疗机器人公司。总部大楼建筑是在一个农业大棚的基础上建造的，它被设计成一条"街道"，周围围绕一系列覆盖木材的建筑体量，这些建筑体量内容纳了色彩丰富的会议空间。"街道"是个多功能空间，既是通道，又是会面和聚会空间，还可以举办各种活动，两端分别是分层的座椅和一面攀岩墙。大面积的玻璃能让人看到装配的过程，并与开放式工作区产生较强的联系。

CMR医疗设备公司是一家位于剑桥的医疗设备初创公司，成立于2014年，目前公司规模已迅速发展到遍布四大洲，拥有300多名员工。

WMB工作室接受委托，为位于剑桥边缘的新跨国公司总部制订计划，将分散在城市各地的员工团队聚集在一起。该项目使公司重新统一，将其革命性的"Versius"医疗机器人系统置于大楼的中心。

该项目场地原先是农业大棚，获得规划许可建造公司总部的基础就是任何新的工业建筑都要复制农业大棚的特殊形式。在因此而设计出的成对的相连门式框架结构中，功能空间是按不同楼层组织的。制造团队、主要的社交空间和访客套房位于一层。开发团队使用的开放式办公桌、会议室和实验室位于二层。建筑的中心是一个双层高的空间，即"街道"，象征性地将所有团队连接在一起。每一台"Versius"机器臂都会在占据空间整个长度的玻璃装配设施内进行最终测试，测试过程就展示在人们眼前。"街道"延伸到通向上层的一组楼梯和阶梯式座椅处，为临时的非正式用途（乃至全公司的聚会）提供了非常灵活的空间。

创始董事会成员和许多员工都是高山登山爱好者。因此，设计任务书中包括在街道的一侧设置一面训练墙。

建筑师在主要室内空间内设计了温暖的带有纹理的材料，目的是与正在开发的精细医疗设备形成鲜明对比，同时起到突显这些医疗设备的作用。二层三个空间的外部均覆盖了桦木胶合板，平行于街道的两面墙体装饰着垂直胶合板，胶合板下面是木棉，这样的设计不仅隔声，在视觉上也很有质感。会议室和茶点间运用了大胆的拼色，方便员工在建筑内寻找方向。

访客套房独立于主要的开发空间和生产空间，因此可以精心策划和控制访客的体验，以吸引投资者、潜在的购买者和参加培训的外科医生。操作空间和街道空间之间的隐蔽连接，使访客可以从系统展示区不知不觉地来到制造和测试过程体验区。

该建筑的主要设计目的是作为CMR的工具，来鼓励以前分散在各地的团队之间产生互动，促进公司未来的发展，并表达公司在医疗设备行业内的颠覆性、动态性。

WMB studio have completed the headquarters building for Cambridge-based CMR Surgical, a leading medical robotics company. Inhabiting an agricultural shed, the scheme is conceived as a “street” surrounded by a series of timber-clad volumes, which house colorful meeting spaces. The “street” is a multi-functional space, used for circulation, meetings, gatherings, and events, anchored at either end by tiered seating and a climbing wall. Large areas of glazing reveal and celebrate the assembly processes, creating a strong connection to the open plan workspaces.

CMR Surgical are a Cambridge-based medical device start-up. Founded in 2014, they have expanded rapidly to over 300 employees across four continents.

二层 first floor

一层 ground floor

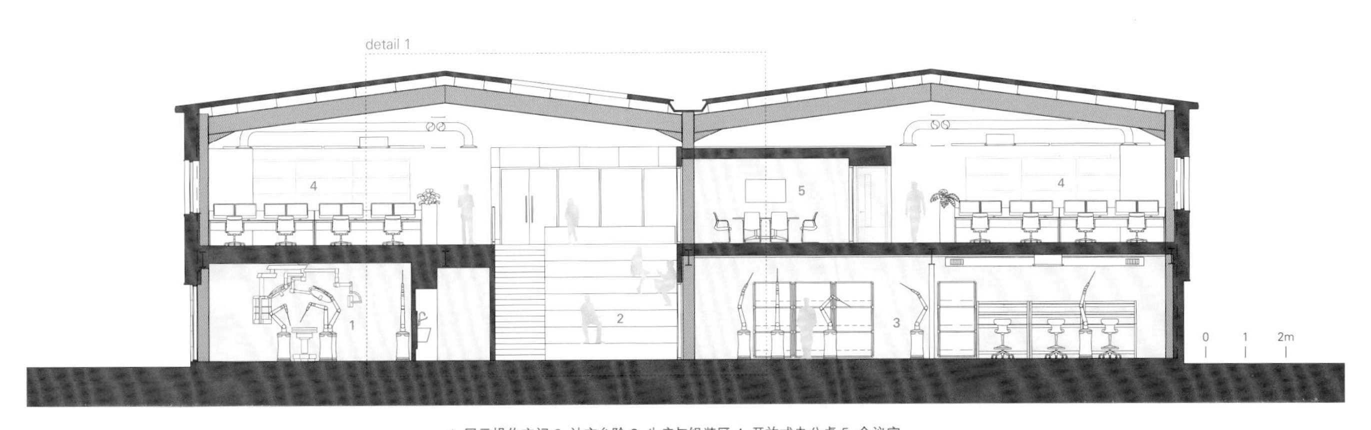

1. 展示操作空间 2. 社交台阶 3. 生产与组装区 4. 开放式办公桌 5. 会议室
1. demonstration operating theater 2. social steps 3. manufacturing and assembly 4. open plan desks 5. meeting room
A-A' 剖面图 section A-A'

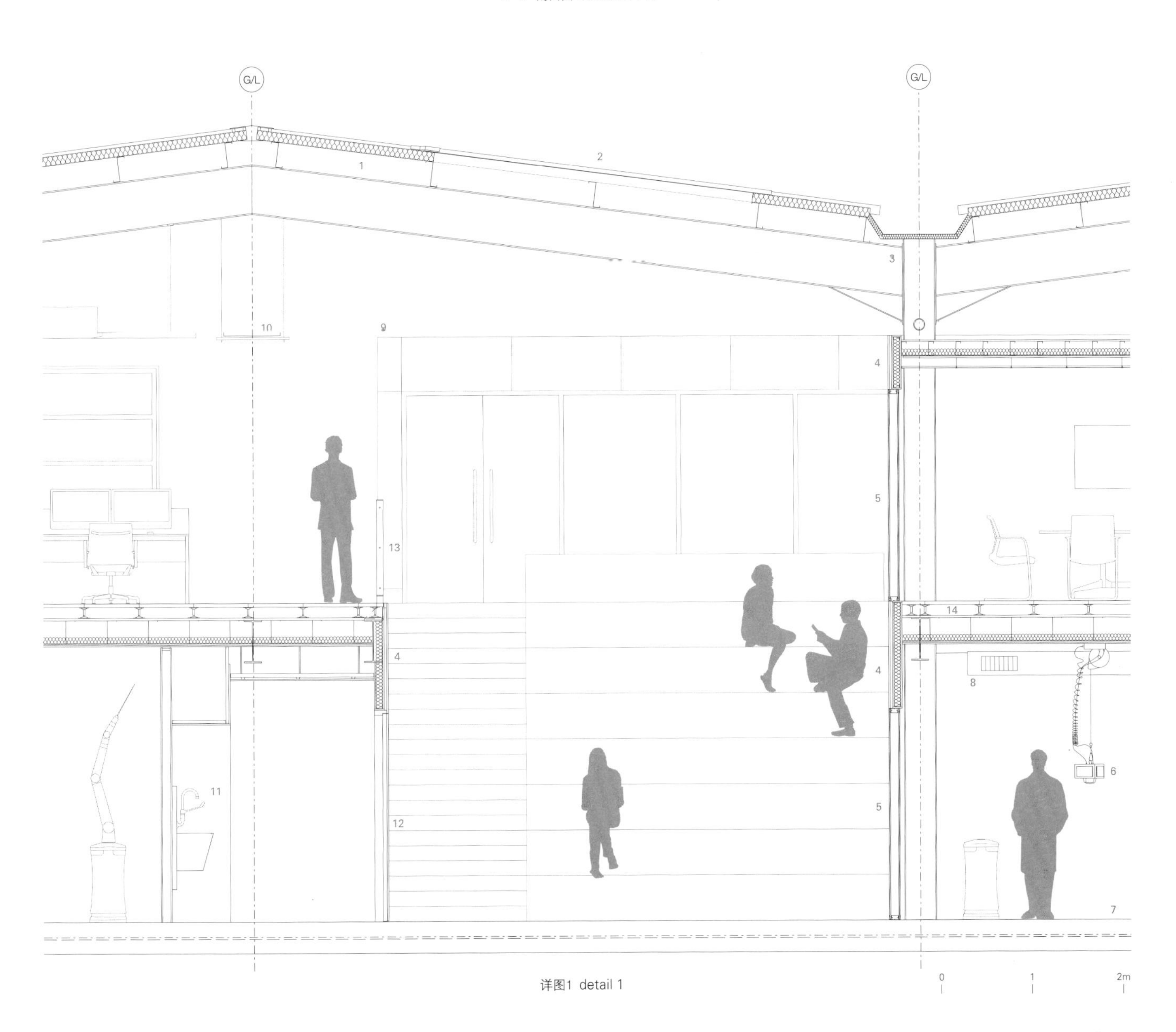

详图1 detail 1

1. insulated composite metal roofing system with perforated underside for acoustic performance 2. industrial roof light 3. primary steel frame, intumescent coated and painted signal yellow (RAL 1003) 4. "street" wall lining; 4-1. 18mm thick birch plywood strips, 50mm wide at 100mm centers, finished with fire protected coating to prevent spread of flame 4-2. 25mm thick wood wool panel, natural finish, superfine 4-3. typical wall build-up; 15mm plasterboard either side of metal stud with acoustic insulation 5. double-glazed interior glazing system flush with outer face of plywood (dark grey frame RAL 7021) 6. suspended, retractable power and data points providing flexible assembly space 7. polyurethane resin floor coating with electrostatic dissipative properties on concrete ground slab 8. suspended ventilation ductwork 9. cladding to first floor enclosures; routed 18mm birch plywood 10. suspended cable tray racetrack 11. scrubs sink for demo operating theatre 12. concealed doorset; door leaf color matched to wood wool and overclad in plywood strips, spaced to match adjacent wall-lining 13. steel balustrade powder-coated dark grey (RAL 7021) 14. raised access floor with zone for data-cabling

项目名称：CMR Surgical HQ Building / 地点：Cambridge, UK / 事务所：WMB studio / 开发商：Cambridge Realty / 开发商建筑师：DT Architects / 客户：CMR Surgical
总楼面面积：3,775m² / 室内装修：Hera Woodwool (25mm thick) "natural" finish; 18mm thick birch faced plywood, used as both routed panels on upper level and in 50mm strips on top of the Hera panels on the ground floor and wall facing the street; colored carpets from Heckmondwike; Remmers poured resin manufacturing/assembly floor (and to street) / 施工时间：2018.4—2019.3 / 摄影师：©French + Tye (courtesy of the architect)

WMB studio were commissioned to develop plans for a new global headquarters on the edge of Cambridge, bringing together staff teams that had been scattered around the city. The project reunifies the company, placing their revolutionary "Versius" surgical robotic system at the heart of the building. Formerly occupied by agricultural shed buildings, the site was granted planning permission on the basis that any new industrial units should replicate these particular forms. Within the resulting pair of connected portal frame structures, functions are organized by level. The manufacturing team, key social spaces and the visitor suite are located at the ground level. Open-plan desks, meeting rooms, and labs, used by the development team, are at the upper level. At the heart of the building is a double-height space, the street, that symbolically connects all teams together. Final testing of each "Versius" arm is expressed and celebrated, within a glazed assembly facility running the length of the space. The street extends into a set of stairs and tiered seating linking to the upper floor, providing flexible space for ad-hoc informal use, right through to full company gatherings.

Founding directors and many of the staff are keen Alpine climbers. Accordingly, a training wall was included in the

brief and is anchored to one side of the street.

For the main interior, a warm, textured material palette was developed, intended to contrast with and exaggerate the finely-honed medical devices being developed. Birch plywood panels clad each of the three enclosures at the first floor. Vertical plywood battens on wood wool dress the parallel walls of the street, providing acoustic damping and visual texture. Bold, block colors are applied to meeting rooms and tea points, acting as orientation points around the building.

The visitor suite operates independently from the main development and manufacturing spaces, allowing the visitor experience, for investors, potential buyers and surgeons attending for training, to be carefully orchestrated and controlled. A concealed connection between the operating theater and the street space allows visitors to move seamlessly from a system demonstration to an experience of the manufacturing and testing processes.

The building is designed primarily as a tool for CMR, to encourage interaction between previously disparate teams, facilitate future growth, and express the disruptive, dynamic nature of the company within the medical device industry.

The Forge办公和展览空间

The Forge

Emrys Architects

伦敦码头区的前米尔沃钢铁厂变成了一个容纳手工艺车间和展览空间的新空间

Former Millwall Ironworks in London's Docklands turns into a new space for craft workshops and exhibitions

艾默里斯建筑师事务所设计的Forge项目是Craft Central的新总部和展览空间。Craft Central是一家致力于推广手工技艺，并为伦敦码头区手工艺专业人士提供经济实惠的工作室空间的慈善机构。该项目被认为是对二级历史保护建筑前米尔沃钢铁厂的一次大胆的现代干预，该钢铁厂最初建于19世纪中叶，当时该地区的造船业达到了巅峰。

设计师将设计和制造业重新引入狗岛区，该地区历来与传统工业有关，但目前由跨国公司主导。Craft Central的目的是通过The Forge项目，在这个艺术家的作品逐渐由于定价过高而无人问津，而且经济也开始下滑的区域支持另一种类型的创意商业模式。

Craft Central与艾默里斯建筑师事务所合作设计了一系列的工作室、工作坊和展览空间。艾默里斯建筑师事务所的设计方案是：在不触及原有建筑的情况下，构建一个独立的两层木结构，保留原建筑的遗产价值和完整性，同时允许在巨大的内部空间内设计新用途空间。新的结构提供一系列独立工作室、会议室，以及位于建筑前部的全高展览和活动空间。

艾默里斯建筑师事务所对工作室所做的设计旨在赋予Craft Central（该机构已在The Forge签订了20年的租用合同）灵活性，使结构模块能适应不同尺寸的单元：独立单元、共享工作室和月租办公桌。工作室面积从7.7m²到26.6m²不等，并且充分利用自然光线。办公桌最大化了The Forge的空间利用率，为Craft Central带来更多额外价值。

建筑师面临的挑战是如何在不影响建筑结构和包括两台高架起重机在内的受保护的内部特色的前提下，使The Forge巨大内部空间的使用更加经济可行。公共区域内的一排看台座位是整个规划中不可或缺的一部分，它为工作室提供了一个醒目的入口，同时也为持牌人提供了一个社交空间。通过看台座位的设计，工作室被从建筑的前部拉回，使这个空间有了更多的用途——不仅可以在手工艺开放日用于展览，还可以用作独立的活动和表演空间。将座位整合到空间布局中，也使人们更接近The Forge的历史结构，同时还能欣赏通过现代干预措施构筑的建筑。

艾默里斯建筑师事务所对该设计任务书的回应出人意料，简洁明快的独立插入结构为建筑增添了额外价值。除了创意性业务外，该项目还为手工作坊、展览和其他社区活动提供了有良好音效的公共设施，自面向公众开放后，已经举办了一场小提琴音乐会。Craft Central现在正在开发该空间的用途，以使其成为当地社区的焦点。建筑前部的大窗户可以让外面的人看到这个空间，从而鼓励路人参与其中。

经过艾默里斯建筑师事务所和iSpace的改造，The Forge的新结构与旧建筑相辅相成。该设计有意将旧建筑的设备和结构暴露出来，进一步彰显主楼的历史传统和特色，而工业材料白桦木胶合板和镀锌钢的使用，则与前钢铁厂的历史建筑语言相得益彰。例如，二层工作室走道胶合板结构的韵律与主楼的柱廊相呼应；而新的介入措施使受保护的高架起重机重获新生，它被放置到入口区域上方，从看台座位上就可以看到。

The Forge by Emrys Architects is a new headquarters and exhibition space for Craft Central, a charity promoting craftsmanship and providing affordable studio space for professionals in London's Docklands. The project has been conceived as a bold contemporary intervention within the historic Grade II listed former Millwall Ironworks – originally built in the mid-19th century at the zenith of the area's ship-building activity.

The project sees design and making reintroduced to the Isle of Dogs district, an area historically associated with traditional industry but currently dominated by global corporations. With The Forge, Craft Central aims to support a different type of creative business model, in an area in which artists have gradually been priced out and a neighbourhood that has witnessed economic decline.

Craft Central approached Emrys Architects to create a collection of studios, workshops and an exhibition space. Emrys's design response constitutes a free-standing two-story timber structure that, by not touching the original building, maintains its heritage value and integrity while permitting new uses within the immense internal space. The structure provides a series of self-contained studios, meeting rooms, and a full height exhibition and event space at the front of the building.

Emrys Architects' design of the studios seeks to grant Craft Central, who have taken on a 20-year lease on The Forge, flexibility with a structural module that can be adapted to allow varying unit sizes: some individual units, some shared studios

项目名称：The Forge / **地点：**London Docklands, UK / **事务所：**Emrys Architects / **承包商、结构工程公司：**iSpace Interiors / **客户：**Craft Central
室内总楼面面积：1,700m² (internal area of ground floor plus the mezzanine deck) / **建筑造价：**£ 1 million / **每平方米的建筑造价：**£ 588.24 per m²
施工时间：2017.5—9 / **摄影师：**©Alan Williams (courtesy of the architect)

SOLEIL
Dry Cleaners
t: 020 7537 4800
SOLEIL Dry Cleaners est. 1988

米尔沃公司内景，米尔沃，1863年
view of the Millwall company's corks, Millwall, 1863

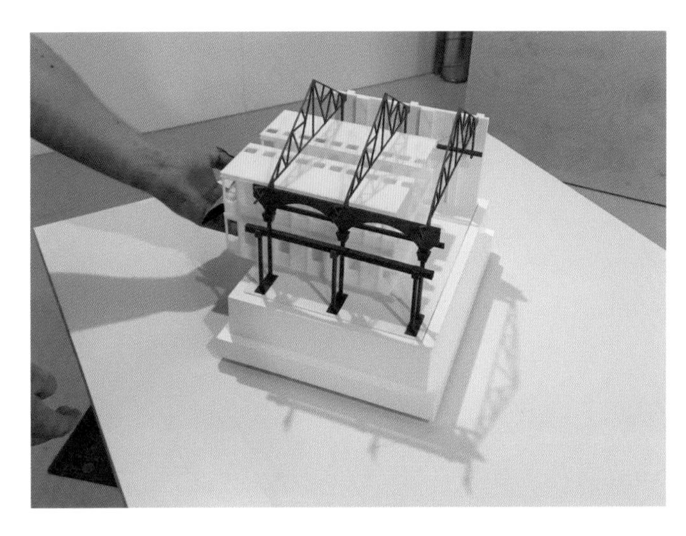

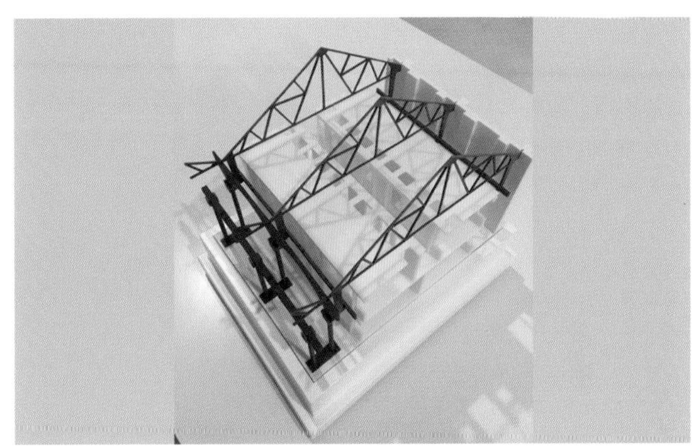

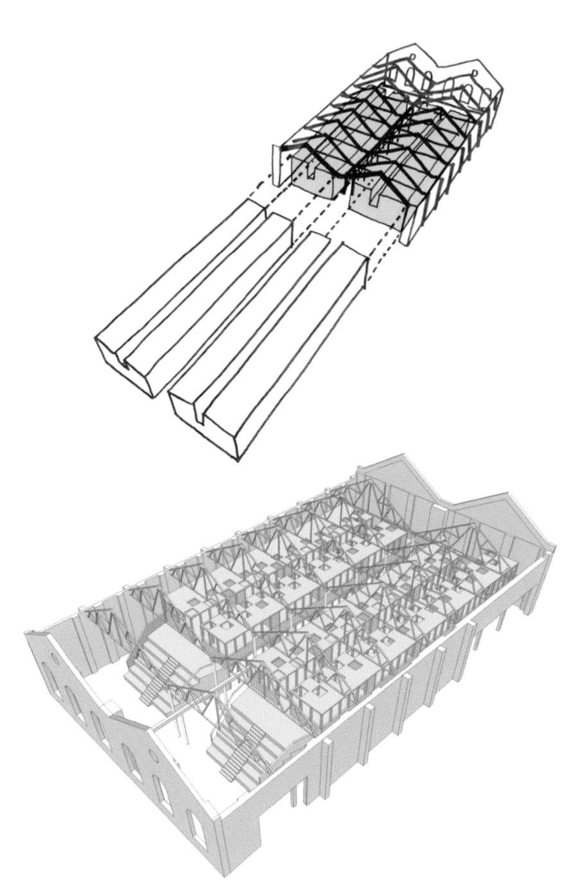

MAYA NJIE
PERFUMES
£15

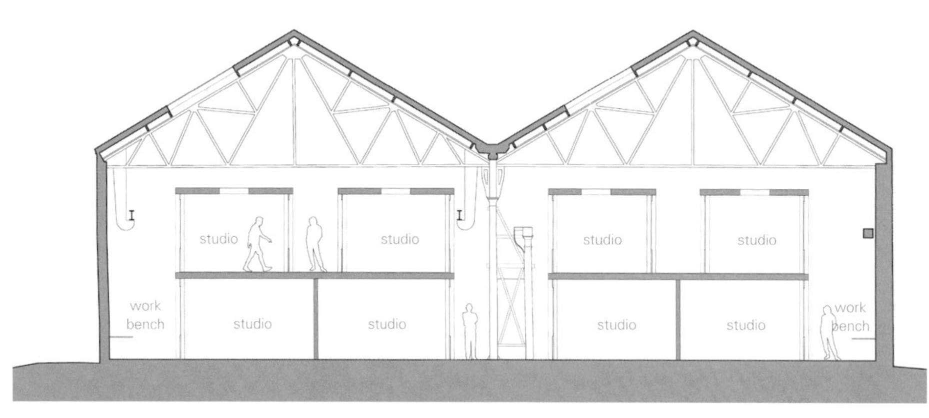

A–A' 剖面图 section A-A'

S.H.HEYWOOD & Co Ltd 4 TONS
Fire exit
SWL 2 TON

and some workdesks for monthly rental. Studios range from 7.7 to 26.6 square meters and make maximum use of natural light. The desks maximize the spatial potential of The Forge and provide added value.

The challenge is to make economic use of the immense internal space of The Forge without compromising the fabric of the building or the internal listed features that include two overhead tracked cranes. A bank of bleacher seating in the public area is integral to unlocking the whole plan, providing a striking entrance to the studios as well as a social space for the licensees. By pulling the studios back from the front of the building and using bleacher seating, the architects created a further use for the space – not only can this void be used for exhibitions on craft open days but also as a separate event and performance space. Integrating the seating into the spatial arrangement also allows people to get much closer to the historical fabric of The Forge, with views of the building framed by the contemporary intervention.

Emrys's surprising yet striking response to the brief, featuring a crisp standalone insertion, has added significant further value to the building. Alongside creative businesses, the project also facilitates a public program of craft workshops, exhibitions and other community-focused activities, facilitated by good acoustics: since opening to the public, the space has already hosted a violin concert. Craft Central are now looking to develop the use of the space and make it a focus for the local community. The large windows at the front of the building look onto this space encouraging interaction from passersby.

The new and the old seek to complement one another throughout Emrys's and iSpace's adaptation of The Forge. A conscious design decision to expose services and structure further celebrates the host building's heritage and character, and the choice of birch plywood and galvanized steel as part of an industrial material palette complements the historical language of the former ironworks. The rhythm of the plywood structure to the upper studio walkways, for example, echoes the column arcade of the main building, while the listed cranes are also brought to life within the new intervention, positioned above the entrance area and viewable from the bleachers.

MANCHESTER
Persil

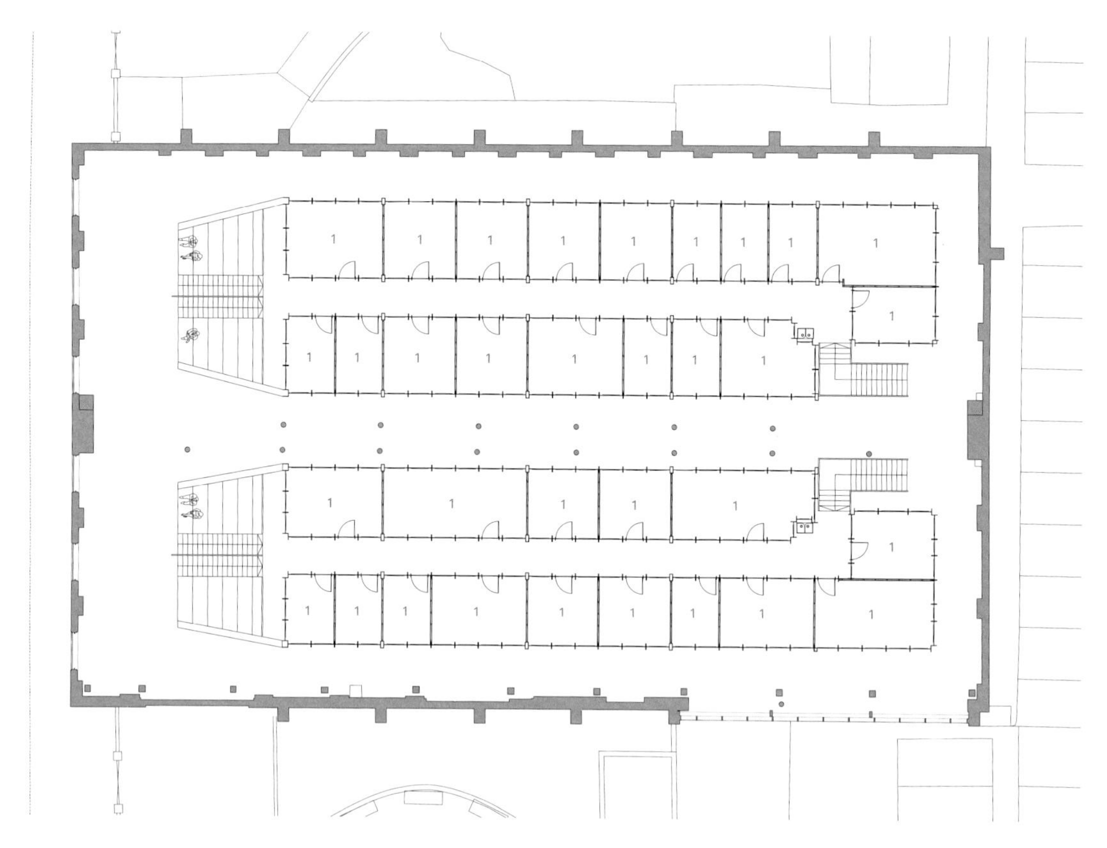

二层 first floor

1. 工作室 2. 工作长椅 3. 野餐长椅 4. 公共厨房区域 5. 男士卫生间 6. 洗手间
7. 新DDA卫生间 8. 女士卫生间 9. 办公室 10. 可移动吧台
1. studio 2. work benches 3. picnic benches 4. communal kitchen area 5. WC's-male
6. wash down area 7. new DDA WC 8. WC's-female 9. office 10. movable bar

一层 ground floor

AMAA阿尔齐尼亚诺分公司办公室
AMAA Branch Office in Arzignano

AMAA

AMAA建筑公司将前工厂改造成自己的现代化办公空间
AMAA adapts a former factory site into their own modern office studio space

建筑就是各个阶段复杂互动的结果，它构成了"项目"。设计不是一个独特的线性过程，而是要不断地审视多种设计方案，始终保持一种批判性的态度。这种实践方式在建筑学中很特别，它基于"概念"在设计过程中的主要角色，同时也基于"操作"的行为，后者也同样重要。它能使建筑过程的三个步骤之间保持平衡，这三个步骤是理论、秩序和最终结果。

每一次迭代都会将原有概念的精髓保留下来，逐渐精确并不断发展，直到项目完成。理论是"概念"和"操作"的综合体：模型、参考资料、文字和其他学科的贡献，都在创作行为中交织在一起。

秩序将这一背景转化为设计，与城市、历史和发展层次建立深层联系。真正的挑战是在从第一幅草图到详细的施工图纸，最后到现场施工完成项目的整个过程中保持住设计概念的根本。一个项目真正完成后，会让原来的设计概念变为现实。

建筑师和客户之间的相互关系是最根本的，比如，卡洛·斯卡帕和安德里尼奥·奥利维蒂之间的合作关系就是证明。共享价值是作品能够有效呈现的基础，也是该工作室建筑研究的完整体现。在这些案例中，客户不仅提出自己的需求和愿望，还投入精力，提供资源，最重要的是，与设计师合作，接受并实现他们的共同愿望。此时此刻，建筑师和客户是一体的。

这种想法的结果就是，在前拱顶工厂内部建造了两层的盒式结构，该结构就是AMAA新的分公司办公室。这座楼中楼在某种程度上与原始的建筑结构融为一体，但同时又是独立的，位于原结构中，与原始废弃仓库的骨架整合在一起。一切都是可见的，包括纤细的钢框架结构、混凝土和金属板、电气系统管线、插头和开关、窗户的金属框架。很明显，建筑非常重视材料的真实性（使用保持原始状态的材料），在成比例模型的制作上也是如此——理论与实践共同组成一个独特的物体。

这些模型就像真实建筑一样，既细致又精确，虽然比例很小，但是却暗示了建筑的复杂性。AMAA追求的是服务于人的空间理念，无论是小细节还是整体设计，始终偏爱定制化的产品。追求风格早已过时，建筑需要一个任何语言都能辨认的脚本，以便对每个位置和任务都能做出精准回应。观察成为协作过程中行动的基础。

Architecture is the result of a complex interaction of phases, making up "the project". Design is not a unique and linear process but is based on a continuous review the multiple options available, always maintaining a critical approach. This manner of practicing is peculiar to architecture; it is based on the primary role of "the idea" in the process and on the act of "doing" which is equally fundamental. It allows a balance to be maintained between the three steps in an architectural process: theory, order and final outcome.

Each iterative step keeps the essence of the original idea and gradually becomes more precise and developes until the project reaches its realization. Theory is a synthesis of "ideas" and "doing": models, references, words and contributions coming from other disciplines, all weaved together in the act of making.

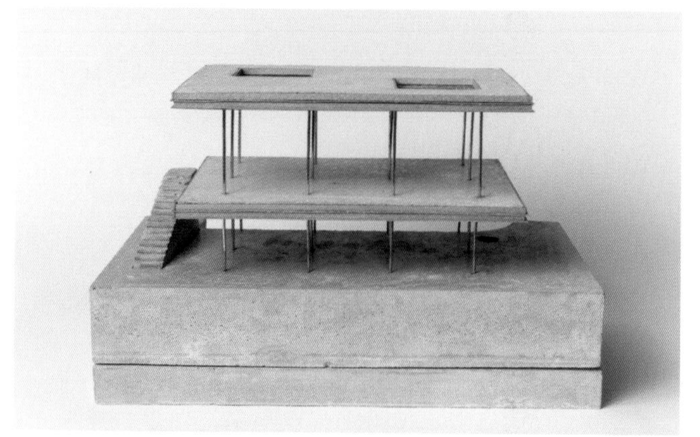

结构框架 structural frame

项目名称：AMAA Branch Office in Arzignano
地点：via L.B. Alberti 11, Arzignano (VI), Italy
事务所：AMAA-collaborative architecture office for research and development
项目团队：Francesca Fasiol
施工公司：main contractor-Il Grifo SRL;
steel frame windows-Santuliana Design;
steel structure-Pettenuzzo Remo S.A.S. Di Pettenuzzo ENRICO & C.
设备：project-Perito Paolo Lucatello;
plumbing-LO.GI.T. Termoidraulica s.n.c. di Antoniazzi Lorenzo e Giovanni;
electrical-ELETTROIMPIANTI di Nogarole Ruggero & C. S.n.c.
客户：Alias SRL-Marco Mettifogo
总楼面面积：180m²
施工时间：2017—2018
摄影师：©Simone Bossi (courtesy of the architect)

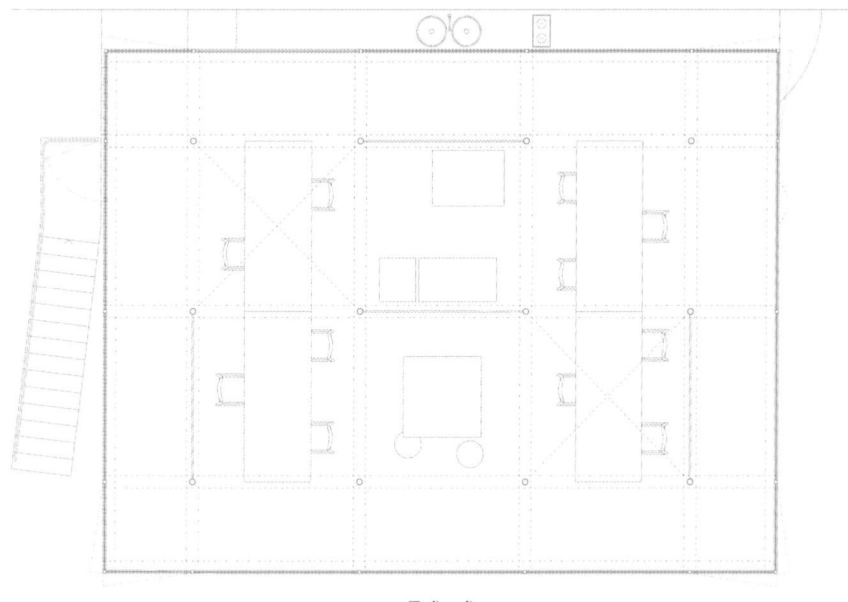

二层 first floor

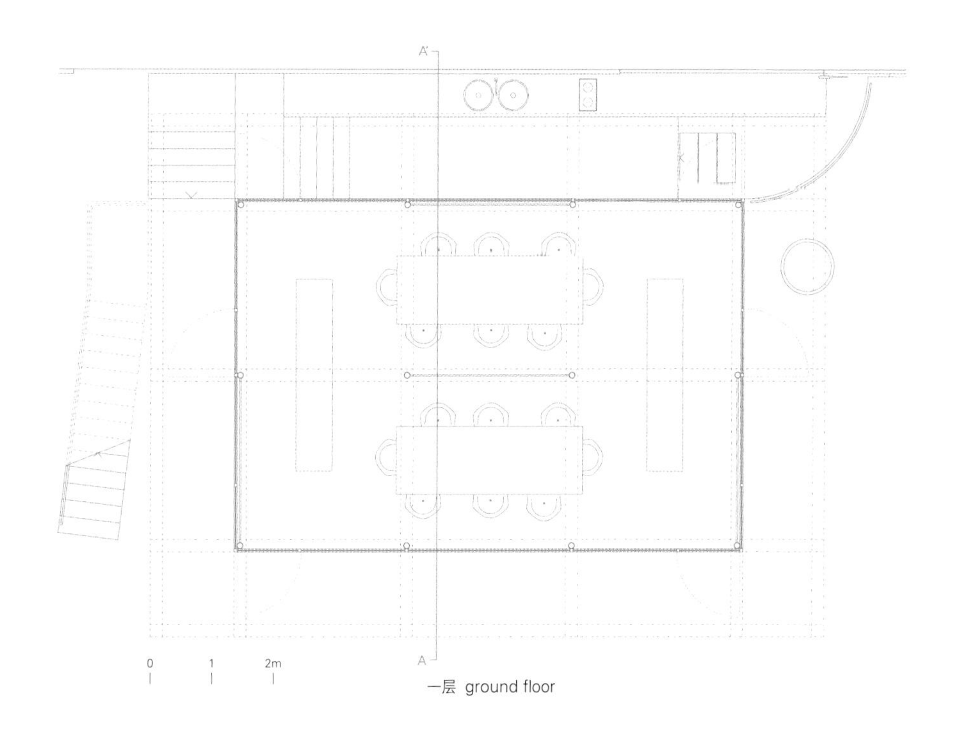

一层 ground floor

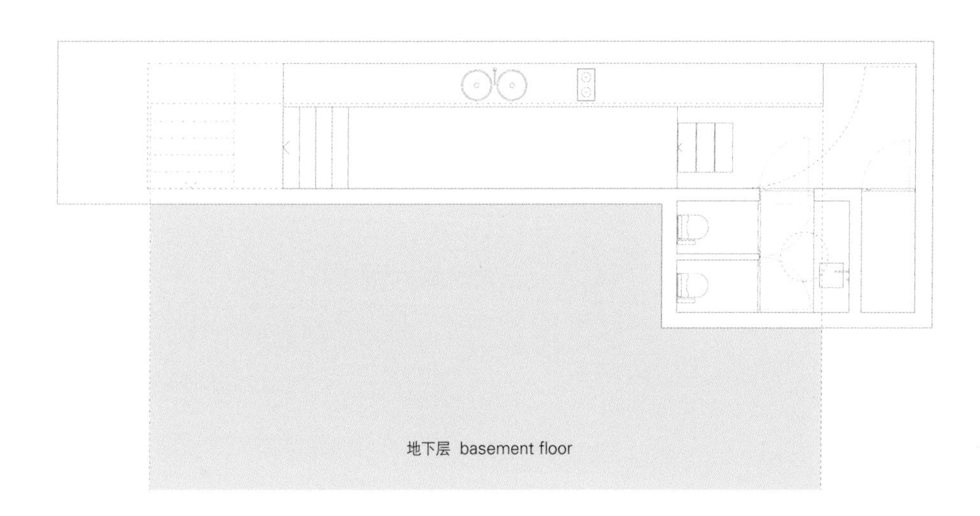

地下层 basement floor

Order transforms this background into a design, forming a deep connection with the city, its history and its layers. The real challenge is to preserve the essence of that idea during the entire process, from the first sketch, to the detailed drawings, and through to the construction on site – the final outcome. The actual realization of a project gives materiality to the original idea.

The reciprocity that binds the architect and the client is fundamental, as already demonstrated by illustrious collaborations and relationships such as, for instance, that between Carlo Scarpa and Adriano Olivetti. Sharing values effectively lays down the foundations for the realization of the work, as a complete expression of the architectural research of the studio. The client, in these cases, not only states their needs or their desires, but puts energy into the process, makes resources available, and above all, works with the designers accepting their shared aspirations. Here, architects and client are one and the same.

The result of such thinking is a two-story box that hosts the new AMAA branch office inside a former vaulted factory: a building within a building, that somehow blends with the original fabric as well as being self-contained, sitting inside, and integrating with, the skeleton of the original derelict warehouse. Everything is visible for what it is: the thin steel frame structure, the concrete and metal slab, the path of the electrical system with its pipes, plugs and switches, the metal frames of the windows. The importance of being truthful with materials (using materials that sincerely reveal themselves) is evident on built architecture but also in the making of scale models, in which theory and praxis come together as a unique object.

These models are as detailed and precise as real architecture; in their small proportions, they give hints of complexity.

AMAA pursues the idea of a space crafted for people, always favoring a custom-made product, from small details to global designs. Style is an outdated phenomenon. Architecture needs a script that is legible in different languages in order to be able to respond adequately to each location and assignment. Observations become the basis for actions, which are encompassed in a collaborative process.

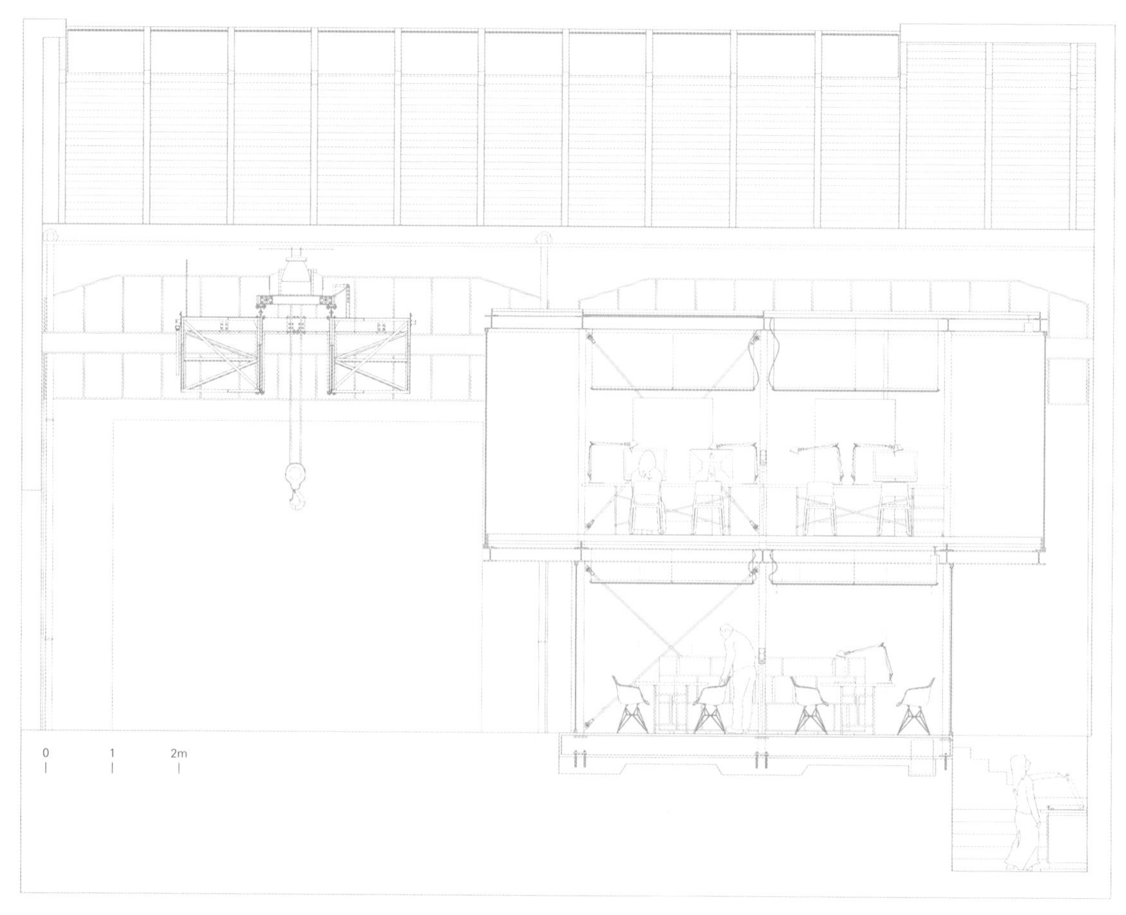

A–A' 剖面图 section A-A'

交织于历史叙事中
在历史背景中建造建筑

Weaving T
Narrative
Building in a H

今天早上，我登上伯尔尼Zytglogge钟楼陡峭的螺旋形楼梯时，我思忖着：这座钟楼的历史意义是什么；几个世纪以来它是如何一直在中世纪城市的中心屹立不倒的；建筑是如何受历史熏陶的，历史是如何与我们身份认同融为一体的。建筑具有连接多个世纪几代人的能力，就像文学一样，它有能力把我们带回到遥远的过去，把祖先的故事讲给我们听。每一座建筑都有自己的故事。显然，历史建筑和它们所叙述的故事对我们而言至关重要。如果历史建筑被破坏，我们会感到愤怒，如果像叙利亚的帕尔米拉这样重要的历史建筑或遗址都被恶意破坏，那就是战争罪。因此，保护代表性历史建筑不仅是道德方面的要求，作为建筑师，我们也有责任帮助保护我们的文化遗产，所以我们会不遗

As I ascended the steep spiral stairs of Bern's Zytglogge clock tower this morning I contemplated the historical significance of the tower and how it had pulsed the heart of the medieval city for centuries. I contemplated how buildings were imbued with history and how our history is integral to our identity. Architecture has the ability to connect generations over centuries of time. Like literature, it has the power to transport us to eras in the distance past and convey stories of our ancestors to us. Each building has its own story to tell. Clearly historic buildings and the narratives they convey are dear to us. If historic buildings are destroyed, we are incensed, and if a significant historic building or site is willfully destroyed, like in Palmyra in Syria, it is regarded as a war crime. Thus there is an ethical aspect to the preservation of exemplary historic architecture and it is our duty as architects to help

ogether of History storic Context

余力地保护我们认为具有内在价值的历史建筑。因此，这就要求我们这些建筑师在历史背景中创建新的建筑干预措施时要高度敏感。在历史遗迹上进行建设，就意味着愿意与复杂的问题和困难阻碍打交道。虽然在历史遗迹上施工可能会让建筑师面临阻碍而头痛，但同时也让他们干劲十足。尽管建筑有严格的建筑法规，但是这些法规有时也会提供解决方案的线索。增加新结构既可以让旧结构脱颖而出，也可能会压制到旧结构的影响力，而新旧结构之间的张力可以产生一种迷人的对话，让二者相得益彰。

preserve our cultural heritage. We go to great lengths to preserve our historic architecture as we deem it to have intrinsic value. It therefore requires great sensibility on the part of us architects to create a new architectural intervention in a historic context. To build on a historic site implies a willingness to engage with complexities and obstacles. Though the hurdles created by working on a historic site might give architects a headache, they also provide impetus. There are stringent building restrictions on listed buildings, but these very restrictions also provide clues to possible solutions. New insertions can suppress or heighten the impact of the old structure and the tension between the old and the new can create a fascinating dialogue that can enhance both.

交织于历史叙事中

在历史背景中建造建筑

Weaving Together Narratives of History

Building in a Historic Context

Anna Roos

本书探讨了当代建筑师如何应对在历史环境中设计新建筑的挑战，着重介绍了建在历史建筑之上、之中或者旁边的新建筑，这些历史建筑中有一些就是废墟。尽管建筑师们的应对措施大相径庭，但所有入选的项目都表现出对历史建筑的高度尊重和高度敏感性，这些新项目或建在历史建筑中，或围绕历史建筑而建。这一部分的项目包括一家建于土耳其考古遗址上的酒店，一个建于奥地利中世纪堡垒内储藏空间中的大型画廊，一个位于丹麦古老城堡遗址旁边的游客中心，一座位于19世纪法国军营中的大学建筑，还有一个位于瑞士的19世纪谷仓改造项目。无论是悬在古老遗址之上，还是嵌入景观中，抑或是环绕着历史房屋，这些分布在全球各地的项目的每一个都展示出建筑师们随机应变的创意方法。这里展示的所有干预措施都让历史建筑继续讲述其迷人的故事，而且不受现代干预措施的影响。

新四星级博物馆酒店（180页）悬在土耳其南部安塔基亚的斯泰利乌斯山考古遗址上，在建筑施工过程中也发现了古代遗址。Emre Arolat建筑师事务所面临的挑战在于如何将酒店与博物馆结合在一起，同时又不让任何一方做出让步。建筑师们没有将遗址掩盖，而是使古老遗迹显露出来，通过将古老遗迹和现代建筑叠加在一起，巧妙地利用考古学和建筑之间的张力。该酒店被构想为四层：一层是露天博物馆，建在考古遗址上方；二层是公共区域、大堂和餐厅；三层是一组组预制的酒店房间，每组之间通过开放的人行道和桥梁相连，游客可以从这里俯瞰美丽的马赛克遗迹；最后是顶层，容纳了舞厅、水疗中心、会议室和餐厅，这一层对建筑模块起到了保护作用。

This book looks at how contemporary architects have responded to the challenge of designing in a historic context; it hones on buildings that are built on, in, or alongside historic structures, some of them mere ruins. Although the architects' responses vary greatly, all the projects chosen show a high degree of deference and sensibility to the historic structures in and around which they are built. Included in this part are a hotel built above an archaeological site in Turkey, a large gallery space inserted within the storage spaces of the medieval fortress in Austria, a visitor center next to an ancient castle ruin in Denmark, a university building in nineteenth century military barracks in France, and a nineteenth century barn conversion in Switzerland. Whether hovering above ancient ruins, embedded in the landscape or encased around a historic house, each of these projects scattered around the globe showcases the various creative ways architects have responded to their particular set of circumstances. All the interventions shown allow the historic buildings to continue telling their fascinating stories without being dominated by the contemporary interventions.
Hovering above ancient ruins of an archaeological site on Mount Starius in Antakya, southern Turkey is the new four-star Museum Hotel (p.180) dating back to antiquity that was revealed during the construction of the building. The challenge of Emre Arolat Architects lay in how to marry a hotel with a museum without compromising either. Rather than concealing the findings, the architects unveil the ancient ruins and play with the tension created between the archaeology and architecture by overlaying the ancient and the modern. The hotel has been conceived in four layers: an open-air museum above the archaeological ruins, the second layer houses the public areas, the lobby and the restaurant above which are a cluster of prefab hotel room pods linked by open walkways and bridges from where visitors can view down onto the beautiful mosaics. The final layer is the canopy housing the ballroom, spa, meeting rooms and restaurant which protects the modules. The structural nodes

安塔基亚博物馆酒店，土耳其
The Museum Hotel Antakya, Turkey

结构节点被小心翼翼地安置在考古层被以前的河流冲刷过的地方，因此马赛克遗迹得以完整地保留了下来。

奥地利维也纳南部的维也纳新城市政府于2016年举办了一场竞赛，旨在开放和重新定义之前隐藏的中世纪堡垒的储藏空间（164页）。这个项目的挑战在于对原有空间进行改造并在壁垒内插入一个大型画廊空间。所以，这里就不能像安塔基亚博物馆酒店那样悬在遗址上面，而应该使新元素与旧元素交织在一起。建筑内部是粉刷成白色的地下空间，这里设有低矮的拱门，经过拱门后会看到突然耸起的高高的皮拉内西式空间。这些空间与崭新的时尚画廊并置在一起，形成鲜明对比。画廊有一半融合在场地中，它是一个独立且精致的覆盖钢板的实体，周围围绕着壁垒。混凝土和精致的波浪钢板与周围厚重的石墙壁垒形成强烈的对比。画廊空间通过弹出式平开窗中的天窗自然采光。景观设计经过精心的坡度处理，将画廊空间与后面的城市公园连接起来，因此地势是从城市向堡垒的历史地下层逐渐降低的，然后又从这里逐渐上升，回到城市和绿化公园的高度水平。

丹麦崎岖的博恩霍尔姆岛上建造了一座哈默斯胡斯新游客中心（194页），位于欧洲北部最大的城堡遗址哈默斯胡斯对面的山谷景观中，是为了接待来到这一著名的历史景点的游客而建造的。和前面奥地利的项目一样，游客中心项目也利用了地势，但在这儿，新建筑与其服务的城堡遗址在地理上是隔开的，而且使遗址位置较低，被谨慎地嵌入岩石地貌中，保持了历史景点的层次感和突出性。实际上，在哈默斯胡斯城堡遗址附近甚至都看不到游客中心。细长的屋顶毫不显眼地在覆盖大片森林的山坡上延伸。建筑师设计了一个缓缓倾斜的屋顶，从而创

have been carefully placed in areas where the archaeological layers were washed away by a former river, thus the architects have left the vestiges of mosaics intact.

The municipality of Wiener Neustadt south of Vienna in Austria held a competition in 2016 to open and redefine the storage spaces of the medieval fortress previously hidden from the public eye (p.164). Here, the challenge was to renovate the existing spaces and insert a large gallery space within the rampart walls. So rather than hovering above the historic layer as in the museum hotel in Antakya, here old and new are intertwined. Within, a warren of subterranean whitewashed spaces with low archways, lofty Piranesi-like spaces suddenly spring skywards. These spaces are juxtaposed with a starkly contrasting new, sleek gallery, half submerged into the site which is pulled away as a separate, delicate steel-clad entity surrounded by the rampart walls. Concrete and delicate steel corrugation strongly contrast with the massive old stone ramparts. The gallery spaces are naturally lit by clerestory lights held aloft in pop-up casements. The landscaping has been carefully sloped to connect the gallery spaces back to the city park, so the topography dips down from the city into the subterranean historical layers of the fortress to emerge back to the level of the city and the green park.

Embedded in the landscape across the valley from northern Europe's largest castle ruin is the new visitor center at Hammershus (p.194) on Denmark's craggy Bornholm island, built to accommodate tourists to the popular historic destination. As with the previous project in Austria, the visitor center also uses the topography, but in this instance by geographically separating the new building from the castle ruin it serves, and positioning it at a lower level, embedded discreetly in the rocky terrain, the hierarchy and prominence of the historic attraction has been maintained. In fact, one can't even see the visitor center when approaching Hammershus. The elongated roof extends unobtrusively

哈默斯胡斯城堡遗址新游客中心，丹麦
New Visitor Center at Hammershus Castle Ruin, Denmark

造了一条架高的人行道，这条人行道是景观的延续，游客可以从这里俯瞰山谷对面的景观和城堡遗址。一条精致的木桥人行道像一条脐带一样，将古老的遗址与新的游客中心连接在一起。在游客中心内，用当地的橡木制成的木配件与厚重的混凝土墙形成了鲜明的对比。Arkitema建筑师事务所将游客中心建造得很低，因此保留了古老遗址的建筑地位。

这里举例的另一个精心设计的嵌入式项目处在城市环境中而非农村环境中，那就是位于巴黎市中心地下的法学院，它位于两座可以追溯至1875年的军事建筑之间（208页）。Chartier Dalix建筑师事务所喜欢把他们创造的现代作品称为"变形"。对角交叉的通道将前阅兵场与地下一层的画廊和演讲厅连接起来。种有树木的小山经过设计师的景观处理，形成了一个设有木凳的户外庭院，法学院的学生们在演讲的间隙可以在此休息。扩建部分掩埋在地下，因此19世纪军营的状态被保留了下来，场地的历史特色也得到了保护。教室位于经过翻修的19世纪建筑中，充分利用了原有建筑的优点——天花板举架高且材料耐久性好。引人入胜的蜿蜒楼梯由黑色钢包裹，将图书馆与二层连接起来。Chartier Dalix试图"调整项目以及新的活动，而不抹杀建筑的过去"。在场地的历史叙事中，建筑师"构思出了源于旧传说的新故事"。

瑞士的Küherhaus（220页）是伯尔尼州它所处时代最大的谷仓，建筑师通过将一个木箱结构小心翼翼地插入已有190年历史的谷仓外壳中，对其进行了翻新。插入原有外壳中的新结构与受保护的结构是分离的，只通过非

across the hillside with the forest rising above it. By slowly ramping the roof, the architects have created a raised walkway as continuation of the landscape from where visitors can view the landscape and castle ruins across the valley. A delicate timber bridge walkway, like an umbilical cord, links the ancient ruins with the new center. Within, timber fittings made with locally sourced oak contrast with massive concrete walls. With their low-slung visitor center, Arkitema Architects have allowed the ancient ruins to retain their status in the architectural hierarchy.

Another discreetly embedded project, in this instance in an urban rather than a rural setting, is the subterranean law university in central Paris, which is buried between a pair of military buildings dating back to 1875 (p.208). Chartier Dalix Architectes like to refer to their modernization as a "metamorphosis". Diagonal criss-crossing pathways link the former parade ground with the subterranean level where the gallery and the lecture theater are situated. A planted hillside is landscaped to create an outdoor forecourt with wooden benches where the law students can enjoy a break between lectures. By burying the extension, the nineteenth century barracks can maintain their stature and the historic character of the site has been preserved. The classrooms are accommodated in the renovated nineteenth century buildings, profiting from their high ceilings and robust materials. A dramatic winding staircase encased in black steel connects the library to the first floor. Chartier Dalix sought to "adapt the project and its new activities without erasing their past." In the historic narrative of the site, the architects have "imagined new stories based on older tales."

Küherhaus in Switzerland (p.220), the largest barn of its era in Canton Bern, has been renovated by the careful insertion of a timber box into the existing 190-year-old shell. The new volume inserted into the existing shell is virtually detached from the protective structure with a minimal of connection

皇家港口园区——Lourcine中心，法国
Campus Port-Royal – Lourcine Center, France

常少的几个连接点相连。Küherhaus位于保存完好的Münchenwiler村的历史中心，该村靠近中世纪小镇Murten，被列入《瑞士联邦国家重要遗产保护名录》。建筑师在遵循瑞士政府严格的历史建筑保护标准的同时，清空了谷仓的内部，并在里面安置了七间不同大小的公寓。垂直布置的公寓的横墙结构与水平屋顶桁架和原有木框架檩条的方向一致。宽敞的房间经过重新布局，采用了最少的干预措施，只配备了像家具一样的独立设施。珍贵的历史建筑构件得到了保留、修复，必要时还使用天然和传统的材料与技术进行了更换和翻新。bernath + widmer建筑师事务所经过精心设计，将这栋美丽的历史谷仓进行了升级和现代化改造，使之成为当代公寓。

通过上述五个项目，我们看到了建筑如何承载记忆，并将不同的时期联系在一起，就像树木的年轮一样，展示了材料的缩影。本书介绍的项目展示了当代建筑师如何使用不同方法克服在历史背景下构建新建筑的困难。无论是将新建筑插入原有结构中，还是与之并列，或悬于其上方，抑或是嵌入其下方，这些都是当代建筑师的应对方式。如果建筑师们能够倾听历史建筑的声音，并对历史内涵做出敏感的回应，他们就能使旧的结构与新的结构交织在一起，并在二者之间创造永不休止的对话关系。在历史背景下建造的建筑不需要受制于历史传承下来的功能和特点，而是可以不断发展，以反映我们这个时代，在尊重自己所处的原有历史结构的前提下，增加新的功能，并以当代的方式对其进行扩建或翻新。因此，历史层次可以在几个世纪的时间里继续为后人增添历史叙事的新篇章。

points linking them. Küherhaus is situated in the well-preserved historic center of Münchenwiler village, which is listed in the Federal Inventory of Swiss Heritage Sites of National Importance near the medieval town, Murten. Whilst complying with the stringent historic preservation standards, the old barn was emptied within and seven apartments of different sizes were inserted inside. The cross walls of the vertically arranged apartments coincide with the horizontal roof trusses and the heights of the purlins of the existing timber frame. The spacious rooms were reorganized with minimal interventions of freestanding, furniture-like installations. Precious historic building elements were preserved, restored and where necessary replaced and refurbished using natural and traditional materials and techniques. With utmost care, bernath + widmer architects have managed to upgrade and modernize the beautiful historic barn to accommodate contemporary apartments.

From the five projects described above, we have seen how buildings carry memories and connect different periods of time revealing a palimpsest of materials like growth rings on a tree. The projects shown in this book illustrate how contemporary architects respond in different ways to the challenges that building in a historic context poses, whether inserted into the old fabric, juxtaposed alongside it, hovering above it, or embedded beneath it. If architects listen to what historic buildings have to say and respond with sensitivity to the historic substance, they can weave the old fabric into the new creating an unending dialogue. Architecture built in a historic context need not be dictated by the function and the identity handed down by history, but can evolve to reflect our current era taking on new functions and being extended or renewed in contemporary ways, while still remaining respectful to the existing historic fabric into which it is built. And so the layers of history can continue adding chapters to their narrative for future generations over centuries of time.

新画廊与炮台
New Gallery and Casemates

Bevk Perović Arhitekti

一个中世纪的弹药库成为奥地利小镇新的公共文化中心

A medieval ammunition storage structure becomes a new public cultural center in an Austrian town

2016年举行的一次竞赛产生了新画廊和炮台项目，目的是将炮台（中世纪堡垒的弹药库结构）改建成向公众开放的新文化中心和展览馆。炮台位于距离奥地利首都维也纳50km的维也纳新城，几个世纪以来，这些炮台一直在不断地被翻新和改建。本案中的这座旧建筑位于市中心和公园的交界处，成为密集的城市结构和开放的城市公园的连接点。

该项目涉及重建和将历史建筑与城市生活融合的问题。设计的核心是"显露"和"添加"——让已隐藏多年的历史综合建筑全部展示在人们眼前，让人们去体验、去理解它，同时增加新的功能。新与旧清晰分离，并相辅相成。

建筑师并没有将原有的构件简单地作为新功能如画般的背景，而是将两者结合在一起，通过新建筑群内外不同时代的空间，建立起一座完全自然的"长廊建筑"。

炮台前的区域被设计成一座坡度平缓的公共广场，从Bahngasse一直延续到炮台的原有高度。广场自然地连接起城市高度和半沉式旧建筑的高度，迎接游客的是一个水平的玻璃结构一层，一层的上方是简单得像盾牌一样的裸露混凝土外墙，外墙的设计参考了原有建筑群的防御功能。进入一层的游客随后被引导着穿过迷宫般的几乎是皮拉内西式的旧炮台结构，这些旧炮台已被改造成城市的展览场所。通过拆除后来加建的结构，原有历史空间的优势得以全部展现。连接大型拱顶储藏空间的一系列地下走廊被清理出来，而不是重建，因此，整座建筑被改造成了一个连续的整体。

地下走廊的尽头是一个新的多功能大厅，大厅从上面采光，有一种艺术馆的氛围。多功能大厅位于维也纳城市公园内，为整个建筑群画上了一个自然又现代的句号。

这个多功能大厅和炮台在同一楼层，而且嵌入地面中，在入口处与历史建筑群相连，这样，历史结构就能作为大型活动的大厅空间使用，可以举办音乐会、展览等。

分别位于多功能大厅和咖啡馆中的两条缓坡将建筑群与城市公园连接起来，从而形成了建筑群的地形轮廓：从城市开始，穿过堡垒的地下历史层，然后回到城市地面和绿色公园。

介入措施的规则通过加建结构的材料定义得到了突显。新加建结构的浇筑混凝土材料补充并显露了历史结构使用的砖材料，新旧建筑明显不同但又和谐统一。

The project for the New Gallery and Casemates resulted from a competition held in 2016, to make the casemates (ammunition storage structures of a medieval fortress) accessible to the general public, in the form of new cultural center and exhibition venue. The casemates, in the small city of Wiener Neustadt, 50km from the Austrian capital Vienna, had been continuously altered and rebuilt over the course of centuries. Positioned on the border between the city center and the park, the old structure became a mediator between the two spatial conditions – that of the dense city fabric, and of the openness of the city park.

The project deals with the issue of reconstruction and integration of historical layers into the life of the city. The operational mode is one of simultaneous "revelation" and "addition" – the historical complex, hidden for a long time, can be experienced and understood in its entirety, while accepting a new programmatic definition. Old and new are clearly distinguished from each other, becoming complementary.

The existing elements are not used simply as a picturesque backdrop for the new functions, but the two are combined in such a way as to establish an entirely natural "promenade architecturale" through the spaces of different eras – both inside and outside the new complex.

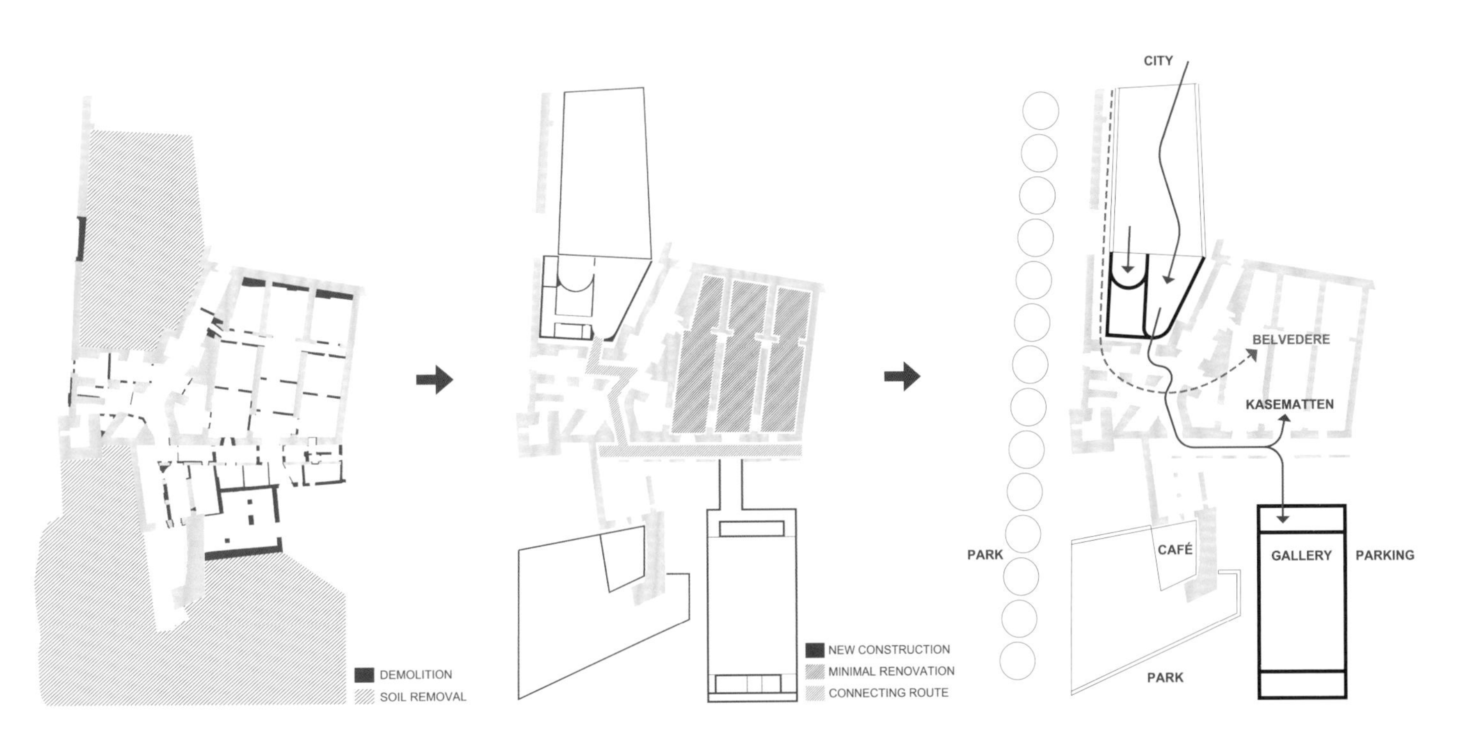
DEMOLITION
SOIL REMOVAL
NEW CONSTRUCTION
MINIMAL RENOVATION
CONNECTING ROUTE
CITY
BELVEDERE
KASEMATTEN
PARK
CAFÉ
GALLERY
PARKING
PARK

WELT
IN BEWEGUNG!
SHOP

The area in front of the casemates is conceived as a gently sloping public square that leads from Bahngasse to the existing level of the casemates. It naturally connects the level of city to the semi-submerged level of the old structure, receiving visitors with a horizontally glazed ground floor and, above it, a simple, shield-like wall of exposed concrete. This is a reference to the defensive function of the existing complex. Visitors are then led through the maze-like, almost "piranesian"structure of the old casemates, that have been transformed into an exhibition venue for the city. Through the process of removal of later additions, the historical spaces are revealed in all their former spatial strength. A series of underground corridors connecting the large vaulted storage spaces has been cleaned, rather than rebuilt, and consequently transformed into a coherent whole.
A new multipurpose hall, a kind of Kunsthalle space, lit from above, sits at the end of the subterranean promenade. It is positioned in the Stadtpark, and provides a natural, contem-

porary conclusion of the complex.
Embedded in the ground, on the same level with the casemates, this multipurpose hall is connected to the historical complex at the entrance area, allowing the historical structure to be used as a lobby space for large events such as concerts and exhibitions.
Two gentle slopes, one in the multipurpose hall, the other in the cafe, connect the space of the casemates back to the level of the Stadtpark, thus finishing the topographical outline of the complex: from the city, through the subterranean historical layers of the fortress, back to the surface of the city and the green park.
The discipline of intervention is stressed though material definition of additions. The brick world of the historical structure is complemented and revealed through the cast concrete materiality of new additions, making the new and the old visibly different but united.

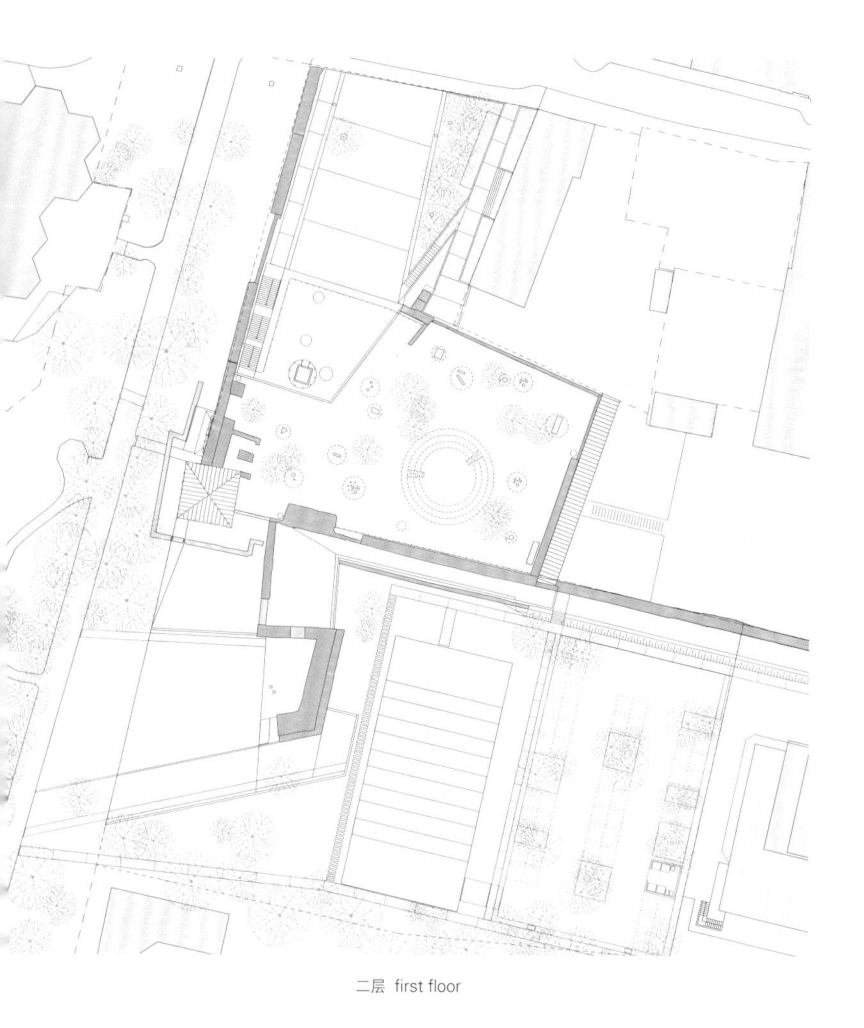

二层 first floor

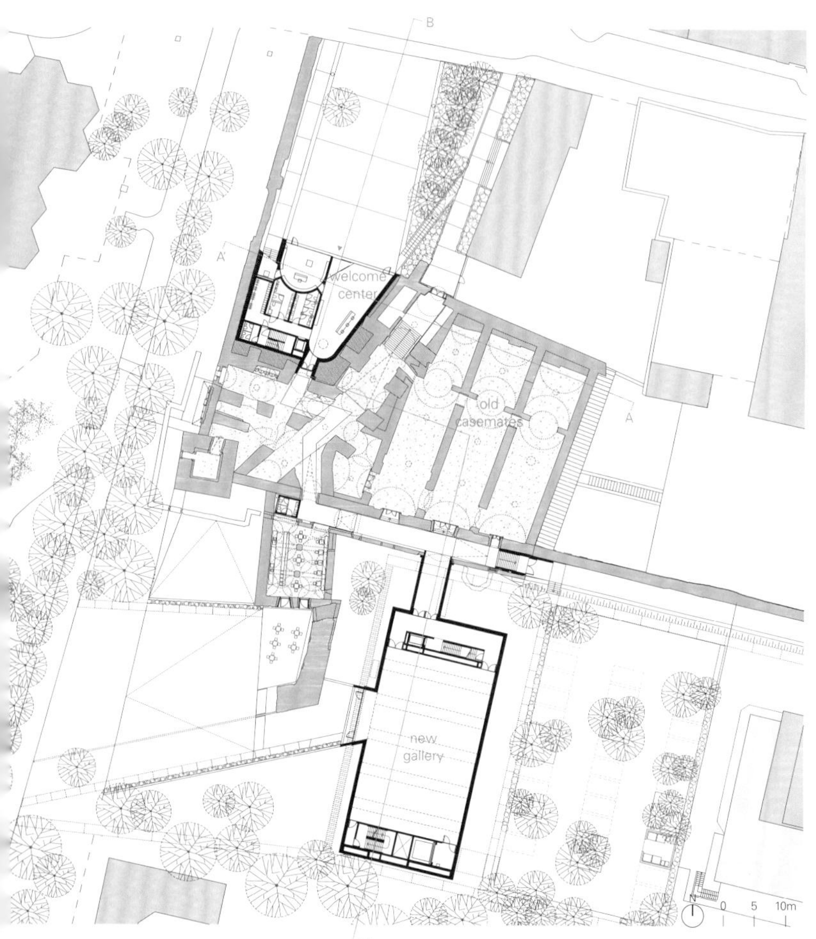

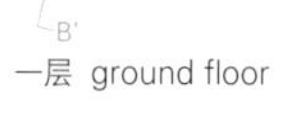

一层 ground floor

ELLUNG
TION

A–A' 剖面图 section A-A'

B-B' 剖面图 section B-B'

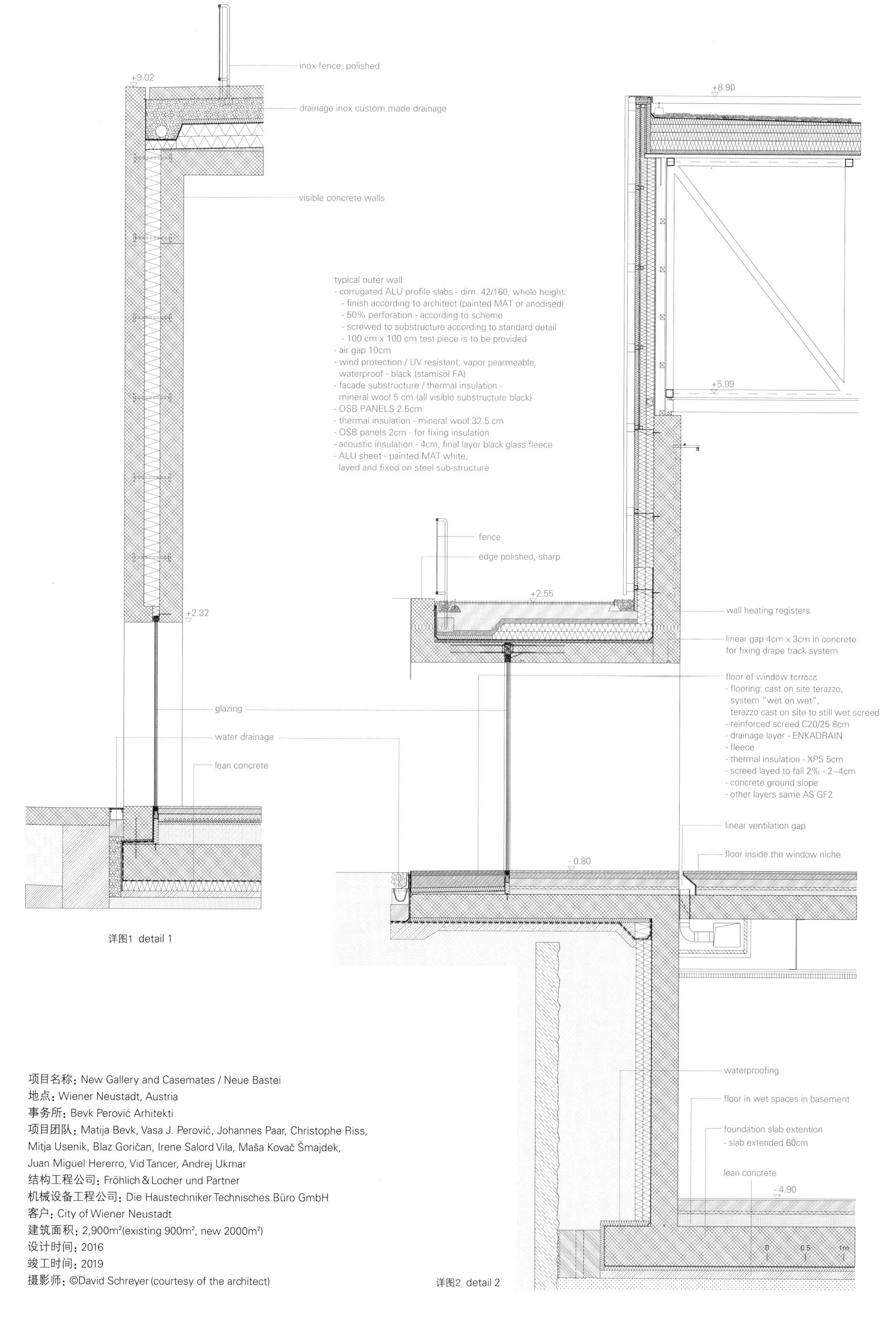

详图1 detail 1

项目名称：New Gallery and Casemates / Neue Bastei
地点：Wiener Neustadt, Austria
事务所：Bevk Perović Arhitekti
项目团队：Matija Bevk, Vasa J. Perović, Johannes Paar, Christophe Riss, Mitja Usenik, Blaz Goričan, Irene Salord Vila, Maša Kovač Šmajdek, Juan Miguel Hererro, Vid Tancer, Andrej Ukmar
结构工程公司：Fröhlich & Locher und Partner
机械设备工程公司：Die Haustechniker Technisches Büro GmbH
客户：City of Wiener Neustadt
建筑面积：2,900m²(existing 900m², new 2000m²)
设计时间：2016
竣工时间：2019
摄影师：©David Schreyer (courtesy of the architect)

详图2 detail 2

安塔基亚博物馆酒店
The Museum Hotel Antakya

EAA – Emre Arolat Architecture

既保留了古老的考古遗迹，又使其融入现代建筑中的土耳其模块化酒店

A modular hotel in Turkey both preserves and integrates ancient archaeological ruins into its architecture

博物馆酒店位于安塔基亚斯塔纽斯山上的圣彼得教堂附近，坐落在一个考古遗址上，据称那里是土耳其第一批基督教会聚集地，可以追溯到古代。这家拥有199间客房的酒店的设计向考古遗址上发现的惊人的马赛克图案、浴池和广场表达了敬意，并且通过将古今交织在一起，关注考古学和建筑之间的紧张关系。

该项目十分重要的一部分就是对独特背景的理解。我们之所以能获得遗产委员会的批准并设计出适合这个地方的建筑，得益于我们熟悉考古保护、重新诠释了酒店类型的建筑方案、利用模块化建筑的专业知识。建筑结构以及更重要的结构点都根据场地的特定区域进行了调整，建筑场地原来是河床，因此考古层早已被冲刷掉了。66根直径为120cm的复合柱连成一个钢格架，将酒店客房和公共空间抬离地面。

建筑分为四层：最靠近遗址的一层是一个露天博物馆；二层是酒店的公共区域、大堂和餐厅，在餐厅可以俯瞰考古遗址上的风景；三层包括酒店客房的预制模块群和露天的交通流线，在那里可以随时欣赏马赛克遗迹的绝妙美景；最后的四层既能保护下方的所有空间，还为宴会厅、水疗中心、会议室和高级餐厅提供了场地，这里被设计成亭子结构，还有种植了树木的庭院，这些开放的公共区域与当地的环境联系十分紧密。

该项目从安塔基亚的当地历史中汲取灵感，力求达到两个主要目标：一是在满足酒店功能的同时，以一种全新的方式展示令人叹为观止的文明痕迹；二是在室内设计中使用现代技术和更多的传统材料。

该建筑采用了高效的被动式通风系统，减少了在交通流线空间内安装机械空调的需要。由于没有外立面，空气可以在人行道和房间之间自由流动。一层的玻璃保护墙旨在阻隔当地盛行的大风和灰尘，同时保护考古遗迹的安全。

Situated near Saint Peter's Church on Mount Starius in Antakya, the Museum Hotel sits on an archaeological site claimed to have housed the first Christian congregation in Turkey, dating back to antiquity. The 199-room hotel pays homage to the amazing mosaics, baths and piazzas discovered during the first drills of the site and draws on the tensioned relationship between archaeology and architecture, by intertwining the ancient with the modern.

An understanding of the unique context was a vital part of this project; becoming familiar with archaeological preservation, re-interpreting the architectural program of the hotel typology and making use of modular building expertise, made it possible to design a building that would fit within this place and receive the approval of the Heritage Committee. The structure, and above all, the points of the structure, were adapted according to the specific areas of the site where, as a former riverbed, archaeological layers were washed away. There are 66 composite columns, 120cm in diameter, all in-

0
20
50m

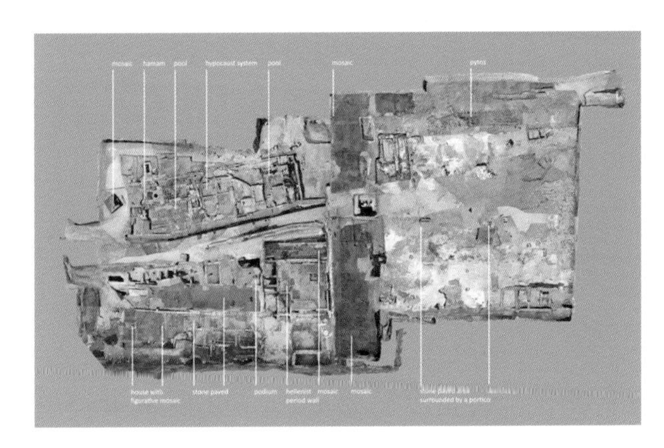
mosaic
hamam
pool
hypocaust system
pool
mosaic
pytos
house with figurative mosaic
stone paved
podium
hellenist period wall
mosaic
mosaic
stone paved area surrounded by a portico

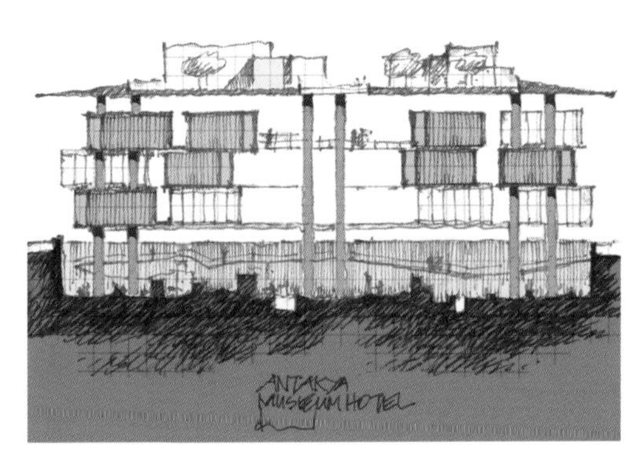
ANTAKYA MUSEUM HOTEL

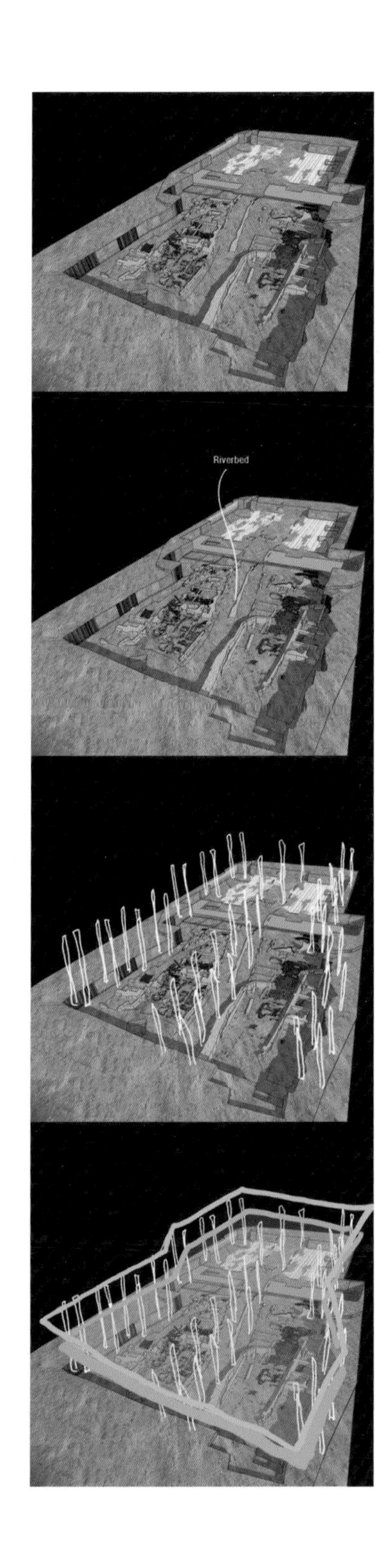

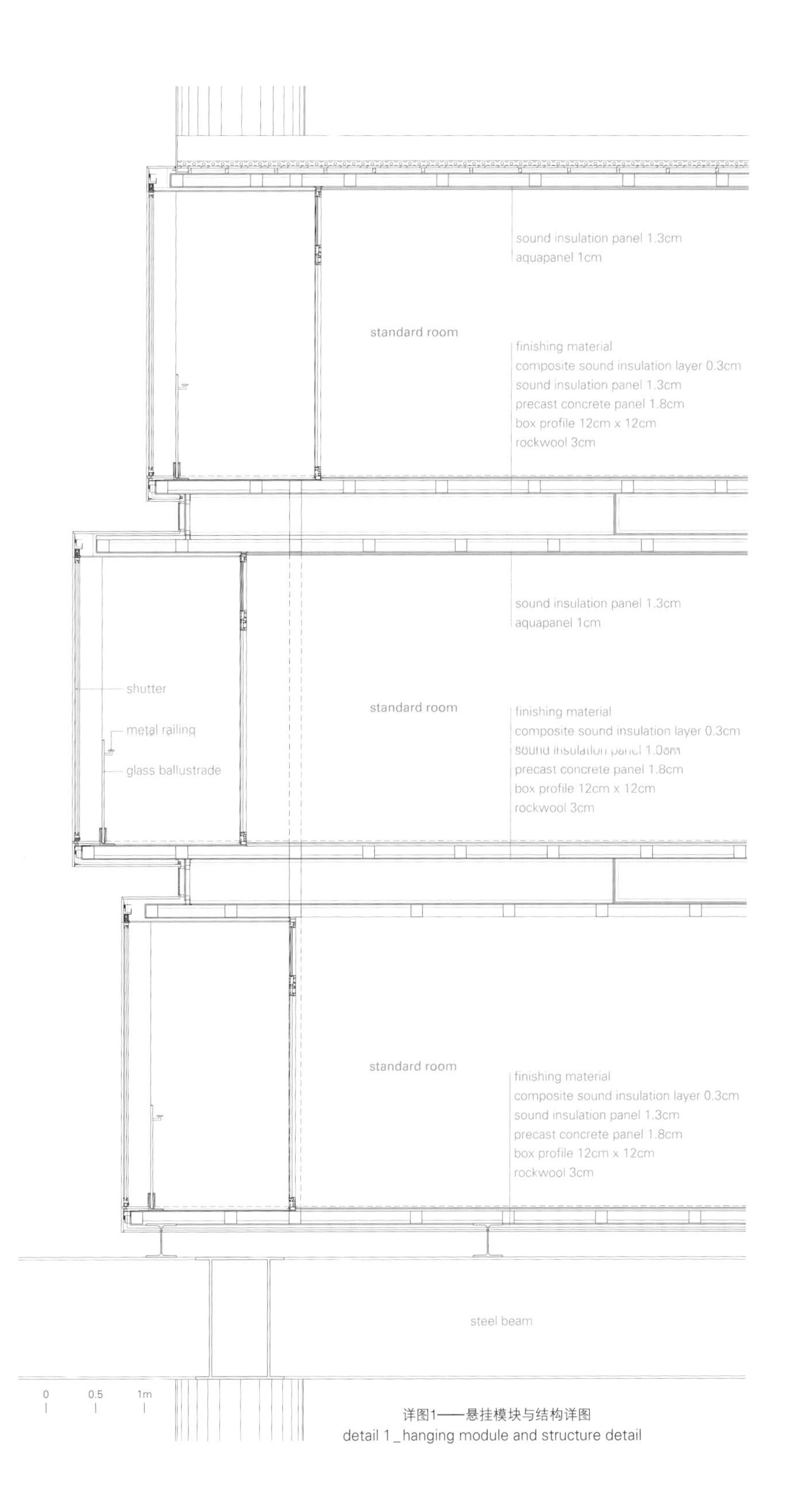

详图1——悬挂模块与结构详图
detail 1_hanging module and structure detail

项目名称：The Museum Hotel Antakya / 地点：Antakya, Turkey / 事务所：EAA-Emre Arolat Architecture / 概念项目团队：leader-Emre Arolat, Başak Akkoyunlu; team-Gülşen Gençalp, Özge Ertoptamış, Rıfat Yılmaz, Anıl Biçer, Tansel Dalgalı, Kaya Sert / Preliminary design project team: leader-Şerif Süveydan; team-Nükhet Ak, Nurgül Yardım, Emre Tunay / IFC项目团队：leader-Neşet Arolat, Gonca Paşolar, Şerif Süveydan; team-Nükhet Ak, Selin Gündüz, Özlem Kayandan, Hüseyin Penbeoğlu / 静态项目：Nodus Mühendislik 机械项目：Birikim Mühendislik / 电气项目：HB Teknik / 照明设计：Piero Castiglioni, SLD Light Design / 防火顾问：Mustafa Özgünler / 声学工程师：Duyal Karagözoğlu / 建筑面积：34,000m² / 设计时间：2014 / 竣工时间：2019 / 摄影师：©Cemal Emden (courtesy of the architect)-p.180~181, p.182, p.183, p.184, p.190[bottom-right] / ©Studio Majo (courtesy of the architect)-p.186~187, p.188~189, p.190[top, bottom-left], p.191, p.192

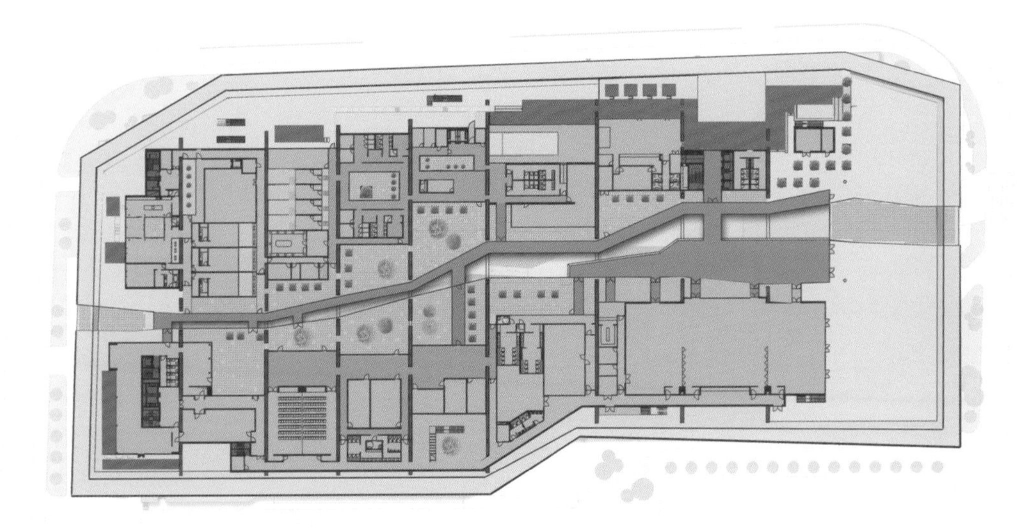

露台层 terrace floor

客房层 rooms level floor

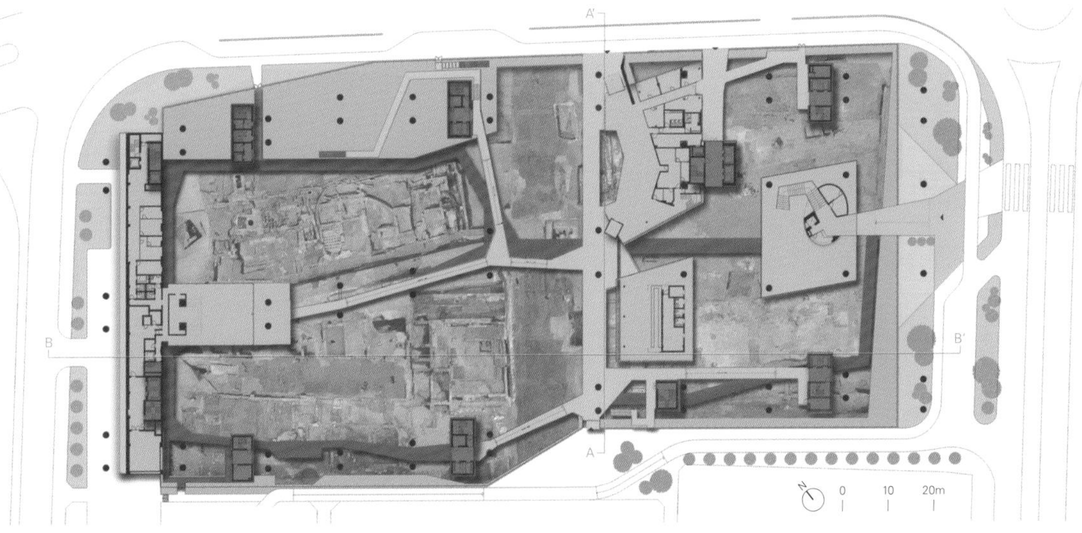

一层 ground floor

terconnected into a steel grid to lift the rooms and common areas of the hotel off the ground.

The building can be considered at four different layers. The first being an open-air museum parkour, at a level closest to the findings. The second layer houses the common public areas of the hotel, the lobby and the restaurant hovering over a scenery of the archaeological findings. The third level is a cluster of prefab modules of hotel rooms and an open-air circulation which keeps retains visibility over the exquisite landscape of mosaics all the time. Lastly, the canopy, which both protects everything underneath and forms the ground for the ballroom, spa, meeting rooms and a specialty restaurant, planned as pavilions and accompanied by tree-filled courtyards – open communal areas with strong ties to the

local context.
Taking inspiration from Antakya's local history, the project sought to meet two main objectives: one was to represent the amazing layers of civilization in an unprecedented way while fulfilling the function of a hotel; and the second was to make use of modern technologies, along with more traditional materials in the interiors.

The building has a highly efficient passive ventilation system, which eliminated the need for mechanical air conditioning in the circulation. Thanks to the omittance of the outer facade, air circulates freely between the footpaths and the rooms. The glass protective wall at the ground level was designed to harness and divert the local prevailing winds and dust, keeping the archeological findings safe.

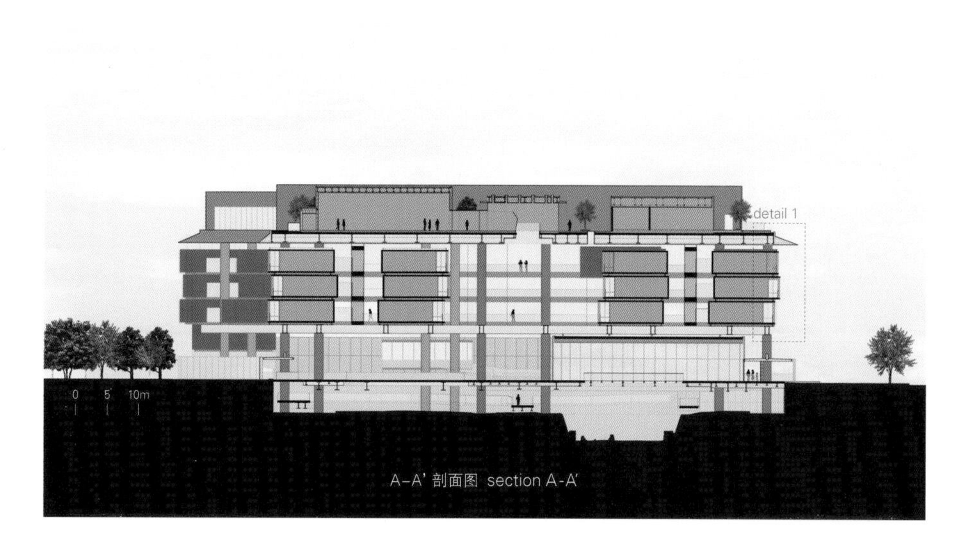

A–A' 剖面图 section A-A'

A–A' 剖面图——透视图
section A-A'_perspective

B–B' 剖面图 section B-B'

哈默斯胡斯城堡遗址新游客中心
New Visitor Center at Hammershus Castle Ruin

Arkitema Architects + Christoffer Harlang

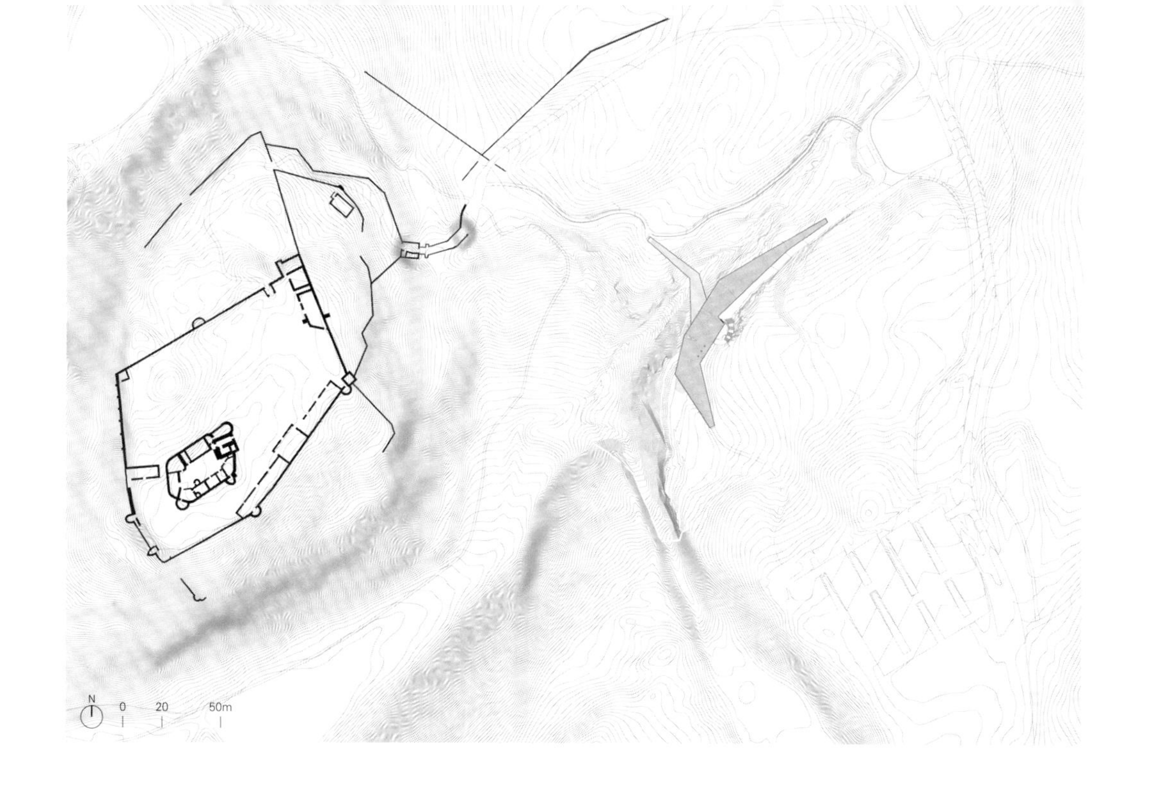
N
0
20
50m

隐藏在自然美景和时间痕迹中的哈默斯胡斯游客中心
Visitor center at Hammerhus hides in the landscape of nature and the traces of time

北欧最大的城堡遗址——丹麦的哈默斯胡斯新建了一个游客中心，由Arkitema建筑师事务所和Christoffer Harlang教授设计，游客中心隐蔽于遗址对面的岩石中。建筑师们的任务是在尊重独特自然环境的基础上，创造一个能提供现代设施和讲故事的空间、能直接欣赏主要景点的地方。

哈默斯胡斯位于波罗的海中一个布满岩石的小岛博恩霍尔姆岛上。经过近五年的建设，游客们现在可以一边坐在咖啡馆中一边欣赏哈默斯胡斯及其周围的风景和海景，也可以在屋顶上散步，拥抱大自然。游客中心建造的初衷是尽量不引人注意，以免抢了周围受保护的森林和城堡的风头。正是基于这样的要求，建筑师们必须设计出一座构思精巧、具有较高美学价值的建筑。

游客中心设计采用简洁的线条和高品质的材料，其中博恩霍尔姆当地的橡木起到了举足轻重的作用。橡木板、特制的橡木设施和家具与由未加工混凝土建成的室内外墙体形成对比，创造了一个迷人温馨的结构。结构上方的悬浮式屋顶成为自然景观路径系统的一部分，一座桥从游客中心前方的露台延伸出来，一直延伸到城堡遗址处，创造了更多的观景点。

拥有独特景观的可用屋顶

游客中心的设计元素之一是形成丘陵景观地形一部分的屋顶。这样的设计可以防止游客在走近时注意到游客中心。这样有两个直接的好处：游客中心建筑不会分散人们欣赏主要景点的注意力；屋顶成为一个有利的自然观景点。

轻型结构屋顶使建筑变得开阔，拥有更好的空间品质，光线可以从两边照射进来。屋顶被设计成一个景观看台，并且配有舒适的座位和宽敞的步行空间，是一个全年均可使用的公共空间。

"我们选择了可以使用的屋顶，来优化游客的观景体验，让游客走近中心时注意不到建筑的外观，" Arkitema高级合伙人Poul Schülein解释道。他说："屋顶成为周围道路系统的自然组成部分，而不会干扰到观景体验。我们为游客提供了一个额外的体验，这种体验是他们在'普通的'游客中心里体验不到的。对我们来说，创造一条动态的观景路线非常重要，这样，游客就有很多机会体验自然环境和历史遗迹。"

A new visitor center at Northern Europe's largest castle ruin, Hammershus, Denmark, designed by Arkitema Architects and Professor Christoffer Harlang, is situated discreetly in the rocks facing the ruin. The architects were tasked with respecting the unique natural surroundings and creating a place which offers modern facilities, space for storytelling and direct views of the main attraction.

Hammershus lies on the small rocky island of Bornholm in the Baltic Sea. After nearly five years of construction work, visitors can now sit in a cafe and enjoy the views of Hammershus, the surrounding landscape and the sea, or they can take a walk on the roof and be in the midst of nature. The center is intended as a discreet building which does not draw attention to itself in relation to the surrounding protected forest or the castle. Precisely this requirement made it necessary for the architects to design a well-conceived piece of architecture with high aesthetic value.

The visitor center is designed with simple lines and high-quality materials, including local oak from Bornholm, which plays a prominent role. The architects play on the contrasts

南立面 south elevation

北立面 north elevation

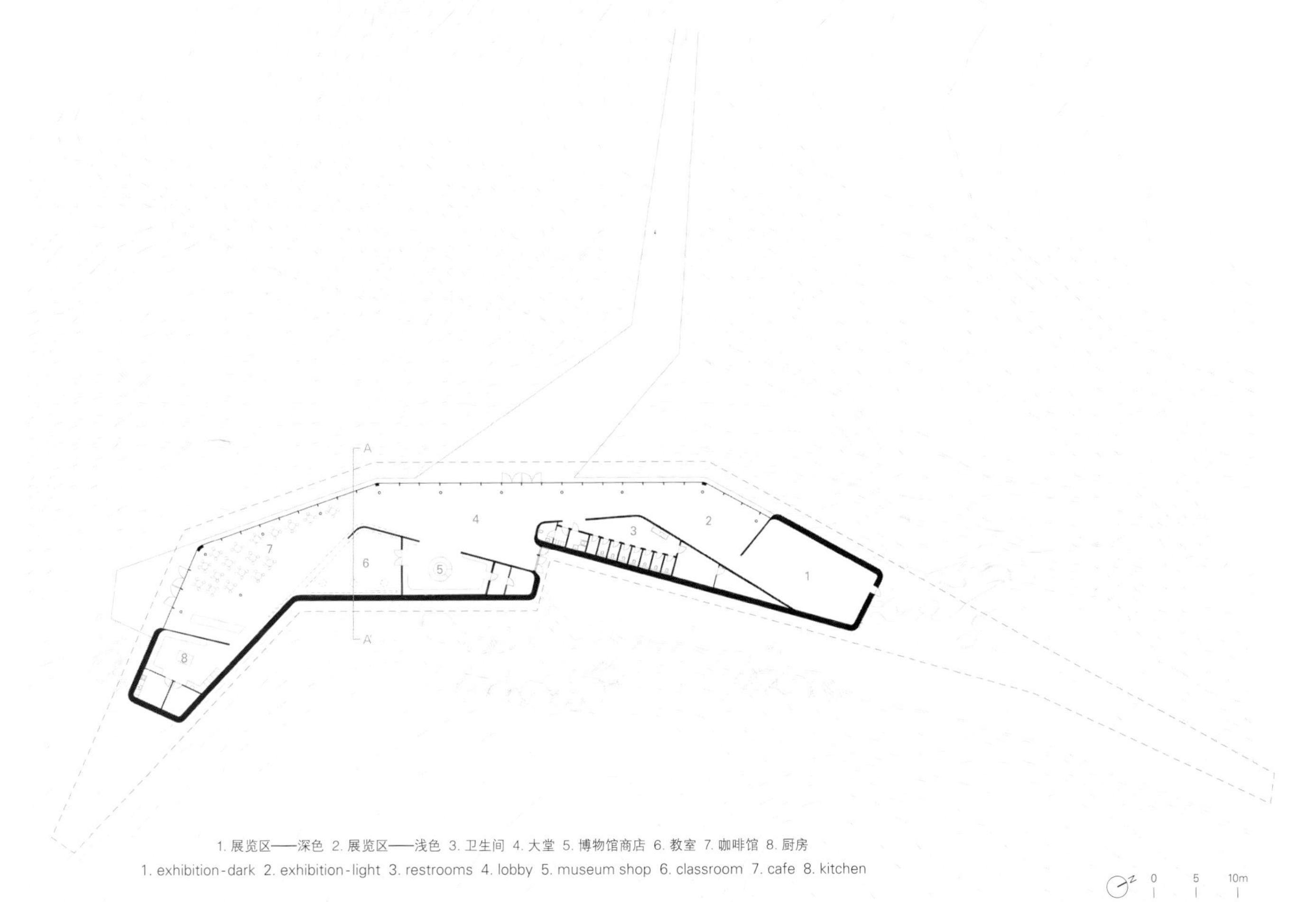

1. 展览区——深色 2. 展览区——浅色 3. 卫生间 4. 大堂 5. 博物馆商店 6. 教室 7. 咖啡馆 8. 厨房
1. exhibition-dark 2. exhibition-light 3. restrooms 4. lobby 5. museum shop 6. classroom 7. cafe 8. kitchen

UDSTILLING
EXHIBITION
AUSSTELLUNG

项目名称：New Visitor Center at Hammershus Castle Ruin / 地点：Hammershus, Bornholm, Denmark / 建筑师：Arkitema Architects + Professor Christoffer Harlang
景观设计：Arkitema Urban Design / 工程公司：Wissenberg / 客户：The Danish Nature Agency / 建筑面积：1,400m² / 设计竞赛时间：2013 / 竣工时间：2018
摄影师：©Jens Lindhe (courtesy of the architect)

of the oak planks, and the specially designed oak fittings and furnishings, with the outer and inner walls of raw concrete to form an inviting and warm structure. Above it, the suspended roof forms a natural part of the scenic path system, and a bridge runs from the terrace in front of the center, continuing the path towards the castle ruin and creating further view-points in the landscape.

Useable roof with unique views

One of the design elements is a roof that forms part of the topography of the hilly landscape. Its design prevents visitors from noticing the visitor center when they approach. This offers two immediate advantages: the building does not distract focus from the main attraction, and the roof becomes a natural vantage point.
This light roof opens out the building, lending it good spatial qualities and light from both sides. It is designed as a land-scape grandstand with good seating and plenty of space for walking about – a public space which is accessible through-out the year.
"We 've chosen a useable roof to highlight the visitors' experience of the landscape and to play down the appear-ance of the building as they approach the center", explains Poul Schülein, Senior Partner at Arkitema. "The roof becomes a natural part of the surrounding path system, rather than a disturbing element in the landscape experience. We give visitors the opportunity for an extra experience that they wouldn't have had if they were merely to walk straight into 'an ordinary' visitor center. It's been important for us to create a dynamic landscape course, in which visitors have many opportunities to experience the natural surroundings and the historical ruin," he said.

A-A' 剖面图
section A-A'

皇家港口园区——Lourcine中心

Campus Port-Royal – Lourcine Center

Chartier Dalix Architectes

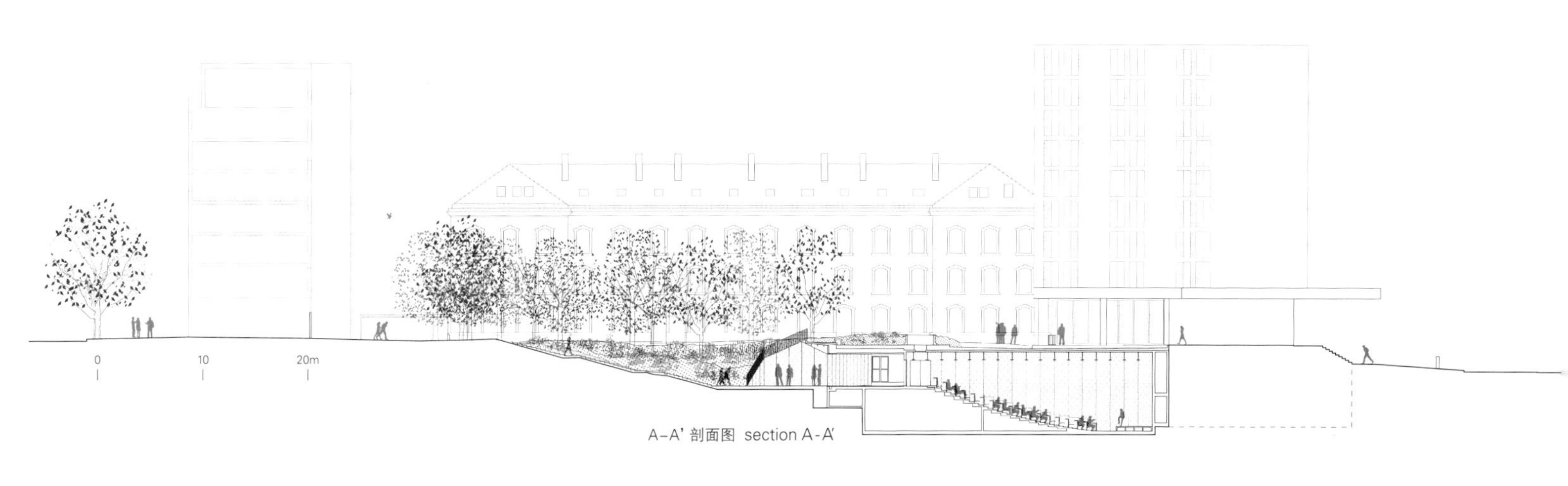

A–A' 剖面图 section A-A'

巴黎军营翻修工程，为巴黎第一大学新建了一座法学院大楼

Renovation of the Lourcine barracks, Paris, creates a new Law Faculty building for Université Paris I

巴黎第13区前Lourcine军营改造是一个雄心勃勃的创新项目。在非常繁华的地区，将兵营向周边环境开放，有助于讲述新故事。采取重新开发的方法不是为了创造一个时间冻结的展览博物馆，而是要让军营以另一种形式再生，即在尽可能保留遗址历史特色的同时，为该区域注入新的生命力。

这个军营在法国大革命前就已经是军用场地，但目前的结构布局可以追溯到19世纪末。阅兵场和皇家港口大道之间是一个地形平坦的区域，朝向冰川河街有一个小坡。不过，由于当时进行了大规模的土方工程和开发，东边的布罗卡街（古老的中世纪小巷）低了约5m。

大型中央阅兵场里种有树木，周围是坚固的营房（1号和2号楼），营房按照古典建筑模式，采用石头、碎石和砖块，结合木框架和石板屋顶，分两期建造完成。

项目场地上迅速建起了办公大楼和宿舍楼，但始终保持着原有的形态和南北主轴线。这两座军事建筑的体量以及大部分的室内布局都得到了保留。

再开发计划包括在1号楼和2号楼以及3号楼的地下层建造巴黎第一大学法学院（教学和研究设施、图书馆和中央复印室）。

3号楼是由法国国防部管理的宿舍，4号楼是让-扎伊中学的学生宿舍。

遗址见证了该地区的城市历史，保护遗址是一项艰巨的任务，尽可能不要改动，但要对军营进行改造才能满足新的功能要求。因此，施工工作是有针对性的，而且仅限于建筑内部；除了改变外门的尺寸以符合现代标准外，围护结构的其他部分几乎没有任何改变。

重建中心广场的战略功能也是关键。不同的用途并存，即大学院系、军区和学生宿舍之间的距离很近，这样人们在利用广场时可能会有冲突，这势必是一个挑战。中心广场必须保持场地的整体感，但又不能让使用者之间不协调发生不愉快。因此，广场不应该是"建造的"，而是恢复的，所以布置接待区尤为重要，既彰显了场地的整体感和凝聚力，又能发挥广场的吸引力、功能性和愉悦性特点。

建筑体现了某种"野兽派风格"，这与公共设施网络外露可见相关，与定制家具的精致细节和原材料（钢、实木橡木、地板）的高贵气质形成对比。建筑采用了自耐候钢材，这是一种温暖、活力无限、多变的材料，从室外公共区域一直延伸到室内，一直陪伴着游客。

图书馆内的过道沿立面布置，光线可以自由照进建筑内，以成排

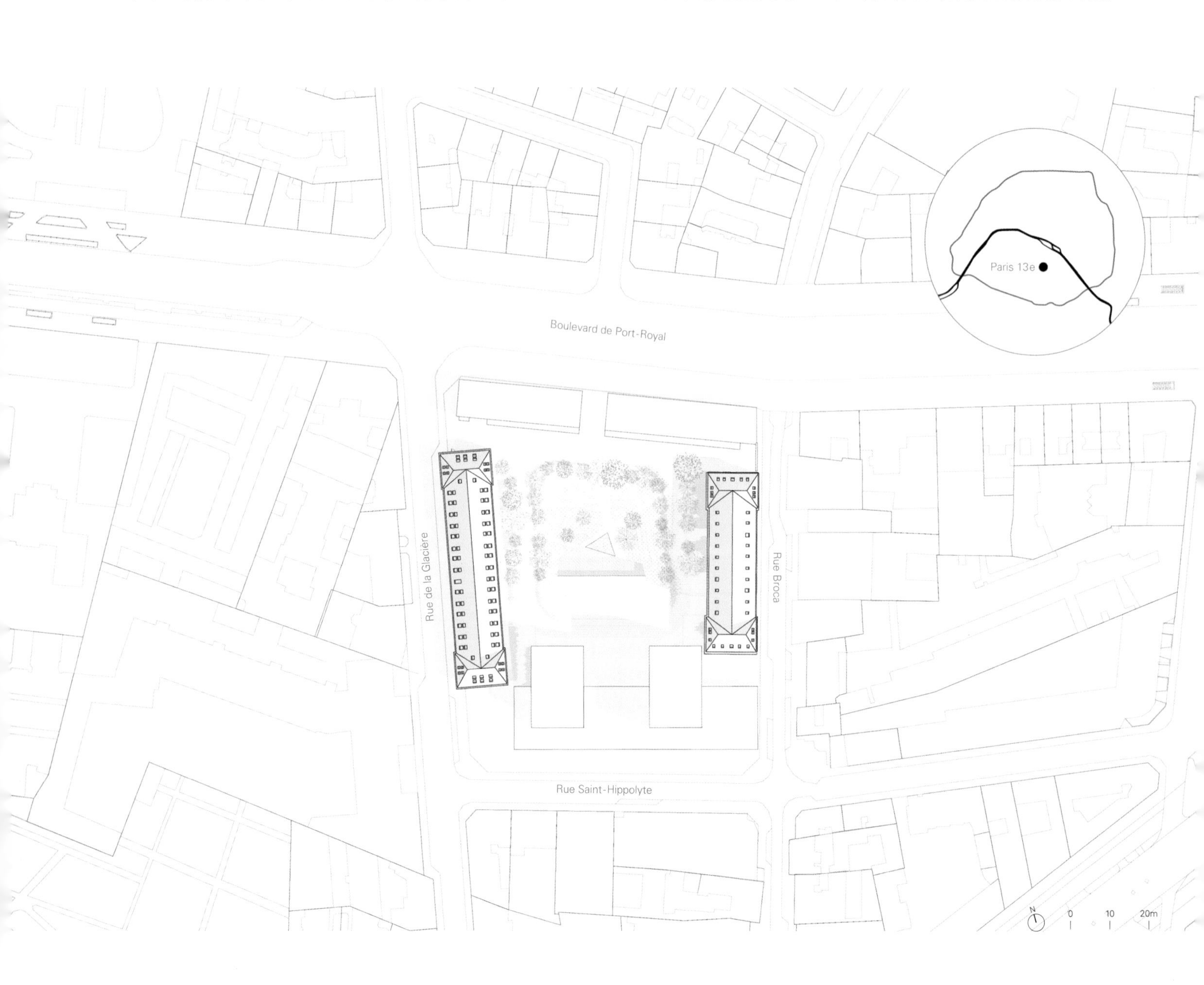

窗户的形式呈现在人们眼前。入口处设有巨大的蜿蜒楼梯。阅兵场的最低点（旧停车场所在的地方）设有一个可容纳500人的露天剧场，剧场入口长廊全部覆盖自耐候钢，与中央广场保持了连续性。

一个花园让学生们可以在欢快的环境中互动。这是一个难得的机会，可以在巴黎市中心建造一座绿树成荫的广场，这样的设计既是对原有梧桐树的展示，也是一种补充，并且提供了一个宽阔的、植被良好的绿色散步场所。

设计的目的是重新发现景观，以便能够讲述其故事，呈现其最好的状态，并且在工作区和生活区之间、建筑空间和开放空间之间提供一个过渡。

The transformation of the Lourcine barracks in Paris' 13th Arrondissement is an ambitious and innovative project. In a very built-up district, opening up the barracks to its immediate environment helps to tell new stories. The approach to the redevelopment was not so much to create a museum exhibit frozen in time, but to regenerate, breathing new life into the district while maintaining its heritage.

The barracks had been a military site since before the French revolution, but the current configuration dates to the late nineteenth century. The topography presents a flat area between the parade ground and Boulevard de Port Royal and a slight slope towards Rue de la Glacière. However on the eastern side, Rue Broca (an ancient medieval lane) is almost five meters lower due to major earthworks and developments undertaken at that time.

A large, central parade ground planted with trees is surrounded by substantial barracks (buildings 1 and 2) which were built in two phases using dressed stones, rubble and brick, with a wooden frame and a slate roof according to a classical architectural model.

The site has evolved considerably, becoming a set of office buildings and accommodation, but always keeping its original form and its major north and south axis. The two military buildings have been preserved both in terms of their volume and a large part of their internal layout.

The redevelopment plans include the installation of the Law Faculty of the Université Paris 1 (teaching and research facilities, a library and central copying room) in buildings 1 and 2, and in the basements under building 3.

002

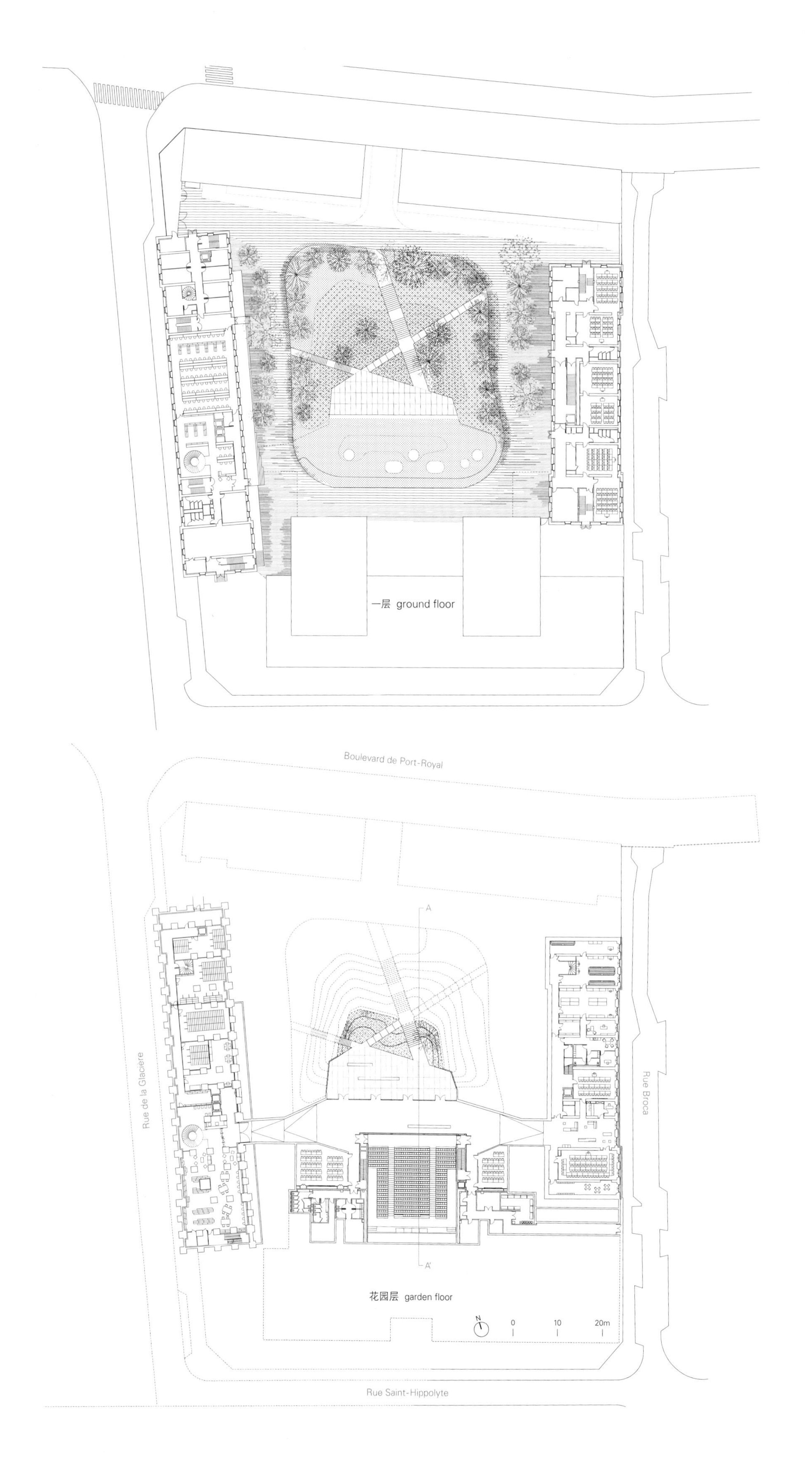
一层 ground floor
Boulevard de Port-Royal
A
Rue de la Glaciere
Rue Broca
A'
花园层 garden floor
N
0
10
20m
Rue Saint-Hippolyte

项目名称：Campus Port Royal-Lourcine Center / 地点：1, Rue de la Glacière, 75013, Paris, France / 建筑师：ChartierDalix / 项目名称：Egis Bâtiment-Tce Elioth-HQE; Acoustb-acoustics; Dhpaysage-landscaping; Grahal-heritage; BTP consultants-supervising office; CSD faces-fire safety coordination BECS-health & safety coordination / 总承包商：Bouygues Bâtiment Ile-de France / 认证：Epaurif, for the Université Paris 1 Panthéon-Sorbonne / 功能：HQE environmental approach (Paris region standard) / 造价：500 seat amphitheatre, 27 classrooms, 2,000m² of library, 1,500m² of offices, 2 official lodgings Gross floor area: 9,710m² / 竣工时间：EUR 22 million / 竣工时间：2019.6 / 摄影师：©Sergio Grazia (courtesy of the architect)-p.208~209, p.212~213, p.215, p.216, p.217, p.218; ©Takuji Shimmura (courtesy of the architect)-p.211; ©Camille Gharbi (courtesy of the architect)-p.219

Building 3 houses accommodation managed by the French Ministry of Defence and building 4 houses students at the Lycée Jean Zay.
The challenge is to preserve the heritage that bears witness to the urban history of the district, altering as little as possible, but refitting the barracks to suit their new intended functions. So work is carefully targeted and limited to the interior; the envelope remains almost untouched apart from altering the size of the external doors to comply with modern standards.
Re-establishing the strategic function of the central square is also key. The co-existence of different uses, namely the proximity between the university faculty, military accommodation and student accommodation, makes the square something of a challenge: it must preserve the overall sense of the site but avoid undesirable confrontations between users. The square should therefore not be "built", but rather restored, so the placement of the reception area is a particularly sensitive feature. This gives the site its overall sense and cohesiveness, playing the part of an attractive, functional and pleasant space.
A certain "brutalism", linked to the visibility of all the utility networks, contrasts with the fine details of the made-to-measure furniture and the nobility of the raw materials (steel, solid oak, floorboards). The self-weathering steel, a warm, vibrant and changing material, accompanies visitors throughout all the exterior public areas, extending into the interior.
In the library, aisles are positioned along the facades to allow light to freely enter the building and present a view of the succession of windows.
A monumental winding staircase marks the entrance.
A 500-seat amphitheater is installed at the lowest point of the parade ground, where the old car park was. It is accessed through a gallery entirely covered with self-weathering steel that provides a continuity with the central esplanade.
A garden allows the students to interact in convivial surroundings. This is a unique opportunity to create a large tree-filled square in the heart of Paris, showcasing and adding to the existing plane trees, to offer a vast, well-planted and green esplanade.
The aim is to rediscover a landscape, to be able to tell its story, show it in its best light and create a space that provides a transition between a place of work and a living area, between a built space and an open space.

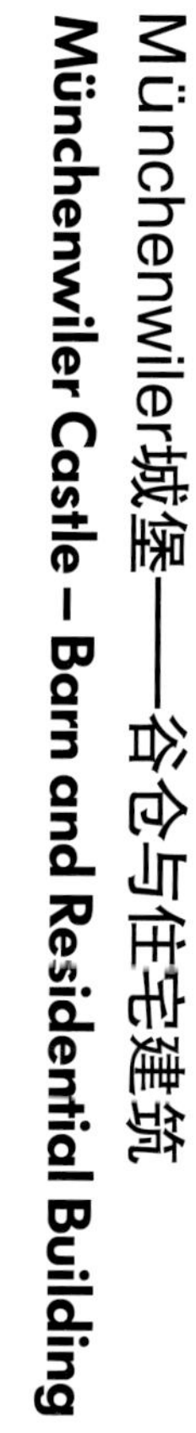

Münchenwiler城堡——谷仓与住宅建筑

Münchenwiler Castle – Barn and Residential Building

bernath + widmer

Münchenwiler城堡一个历史悠久的谷仓在尊重原貌的基础上被改建为住宅建筑
A historic barn in Münchenwiler Castle converted into a residential building while respecting the original

Münchenwiler村保存完好的历史中心位于瑞士中世纪小镇穆尔滕附近，已被列入《瑞士联邦国家重要遗产保护名录》。

这座列管建筑曾经是著名的Münchenwiler城堡庄园的一部分；城堡前身是修道院，至今仍然是小镇内的历史性建筑。该庄园拥有当时伯尔尼州最大的谷仓，是1830年左右在原有的基础上建造的。

该房产已被改造成住宅。雄伟的空谷仓结构中建了7套公寓，所有的建筑工程均符合历史保护标准；插入原有外壳中的新体量实际上与受保护的结构互相分离，它们之间只有很少的连接点，从而使原有的历史结构保持完整。这种策略在新体量和外表皮之间的上层创造了一个有趣的半室内阳台空间。光线可以穿透水平的木板条，用户可以向外眺望（尽管视角有限），但从外向内看却几乎是看不到的，很好地保护了私密性。

前农场建筑的功能多样性体现在了新公寓的类型上。这些公寓垂直排列，并通过砖砌的防火墙彼此隔开。

附属三层公寓的横墙是根据水平屋顶桁架的间距和原有木框架檩条的高度来布置的。公寓入口是住宅楼以前的北入口。南面是花园/庭院；景观建筑由苏黎世的园林绿化专家安德烈亚斯·盖泽尔齐在前农家庭院里完成的。该建筑俯瞰着乡村景观全景和老城区。

与较大谷仓相连的前"Küherhaus"可以追溯到1620年，这座古老的建筑在19世纪时已经进行过翻新。它原来是住宅楼，与谷仓的外壳一起得到精心修复。在修复过程中，住宅楼被分成了两栋单层公寓。宽敞的房间通过最少的干预措施，采用独立式、类似家具的设施进行了重新布局。可逆的轻质干式新楼板结构与更高要求的天花板相契合。项目中使用的材料与原有的材料和谐匹配，例如，云杉、橡木、石材和金属。材料的连续性让老建筑无论在视觉上还是触觉上都与现代加建建筑融为一体。最终，珍贵的历史建筑构件得以保留、修复，必要时还进行了更换和完善。施工中只采用了自然和传统的材料和技术，不仅是对原历史建筑的一种尊重，也是一种补充。

The well-preserved historic center of Münchenwiler village, near the little medieval town of Murten, Switzerland, is listed in the Federal Inventory of Swiss Heritage Sites of National Importance.

The listed building used to be part of the well-known Münchenwiler Castle Estate; the castle, itself a former priory, still stands as a historic building within the town. The estate houses what was the largest barn in the Canton of Bern at the time. This was built on its existing foundations around 1830.

The property has been converted into a residential dwelling. Seven apartments were built into the monumental emptied barn structure. All building work was in compliance with historic preservation standards; the new volume inserted into the existing shell is virtually detached from the protected structure, with only few connection points between them, allowing the original historic fabric to remain intact. This

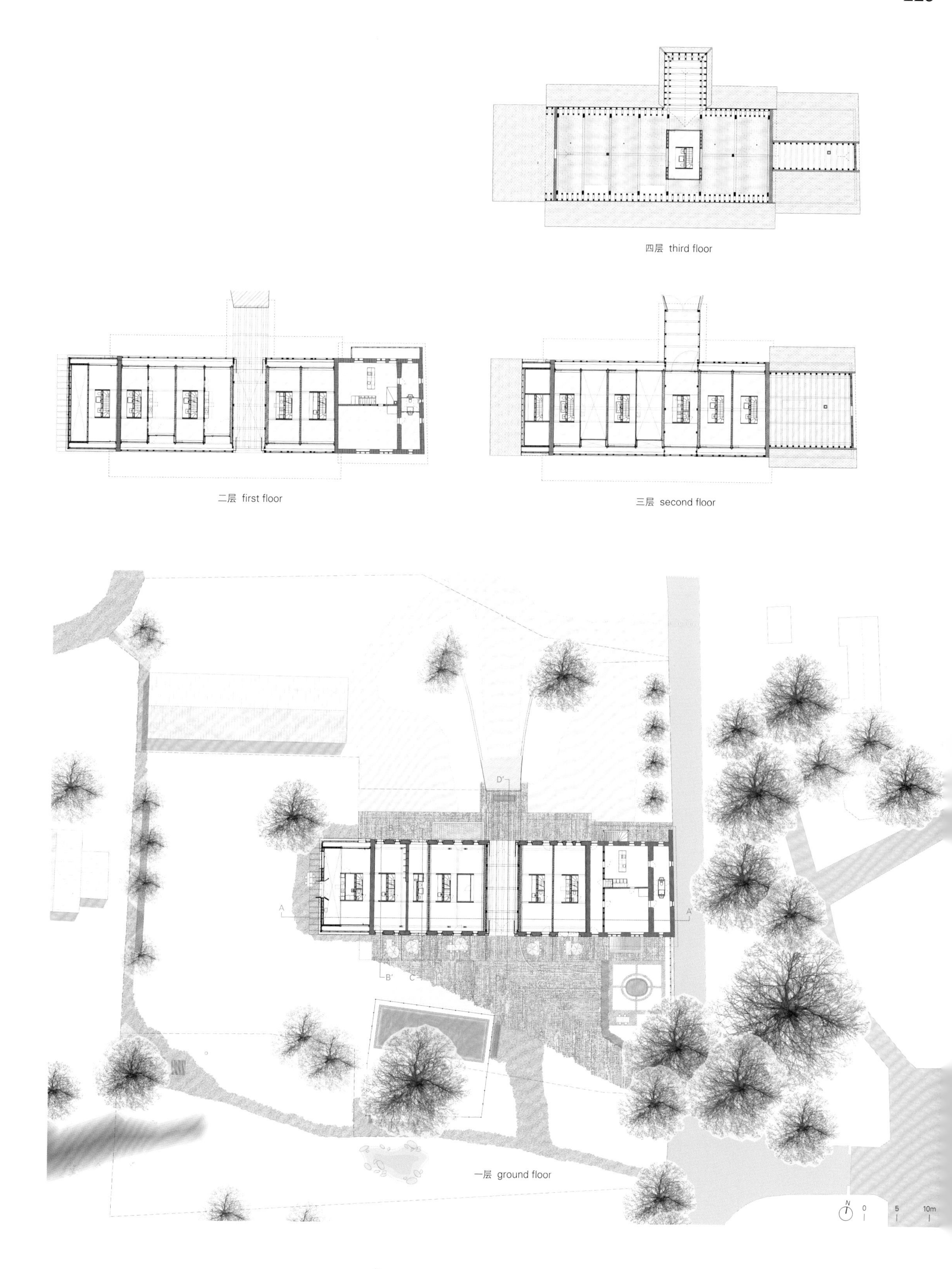

四层 third floor

二层 first floor

三层 second floor

一层 ground floor

strategy creates an intriguing semi-interior balcony space to be achieved on the upper floors, between the new volume and the outer skin. Light permeates through the horizontal wooden slats which allow for (albeit somewhat limited) views outwards and maximum privacy when looking in.
The functional versatility of the former farm building is reflected in the typology of the new apartments. These are arranged in vertical order and separated from each other by brick firewalls.
The cross-walls of the attached three-story apartments are arranged according to the spacing of the horizontal roof trusses and the heights of the purlins of the existing wooden frame.
Entry to the apartments is through original openings in the dwelling, to the north. To the south are the gardens/court-yard; landscape architecture was completed by Zurich-based Andreas Geser on the former farmyard. The property over-looks a panoramic rural landscape and the old town.

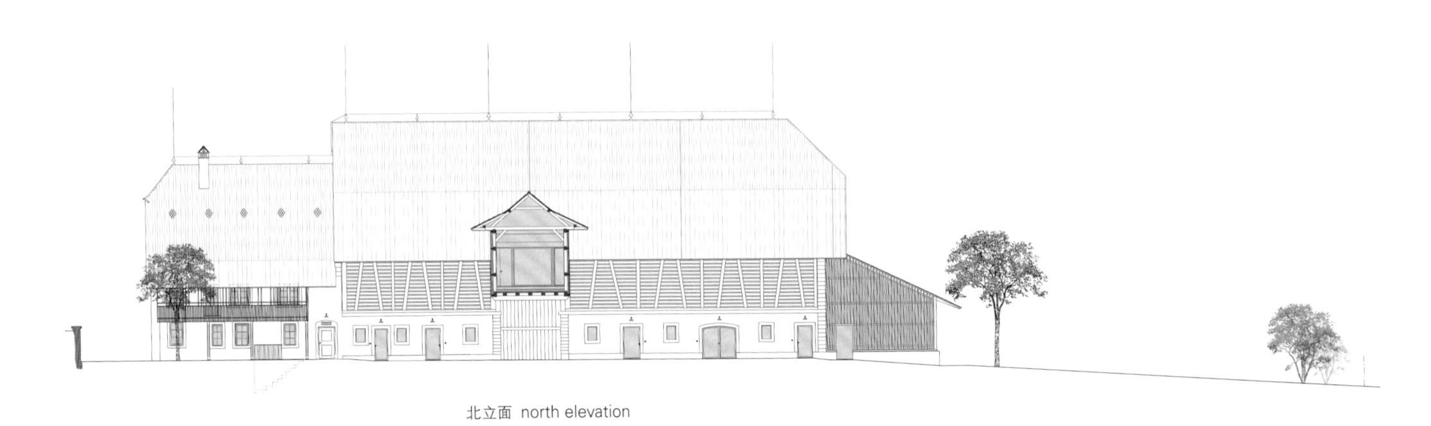

北立面 north elevation

东立面 east elevation

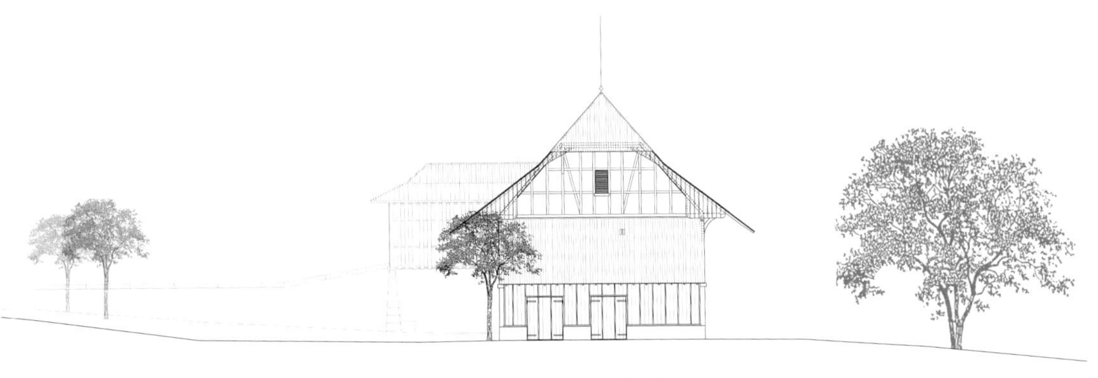

西立面 west elevation

The former "Küherhaus", attached to the larger barn, dates back to 1620. This old structure received some subsequent modifications in the 19th century. This original residential building is carefully restored, together with the outer shell of the barn. In the course of this restoration, the residential building is divided into two single-floor apartments.
The spacious rooms are reorganized with the minimal interventions of freestanding, furniture-like installations. The higher requirements for the ceilings are met by new, reversible, lightweight and dry floor constructions. Materials used in the project match the original material palette: spruce, oak, stone and metal. This continuity of materials joins the old building with the modern additions both haptically and visually. Ultimately the precious historic building elements are preserved, restored, and where necessary, replaced and completed. Only natural and traditional materials and techniques are employed, respecting yet augmenting the original.

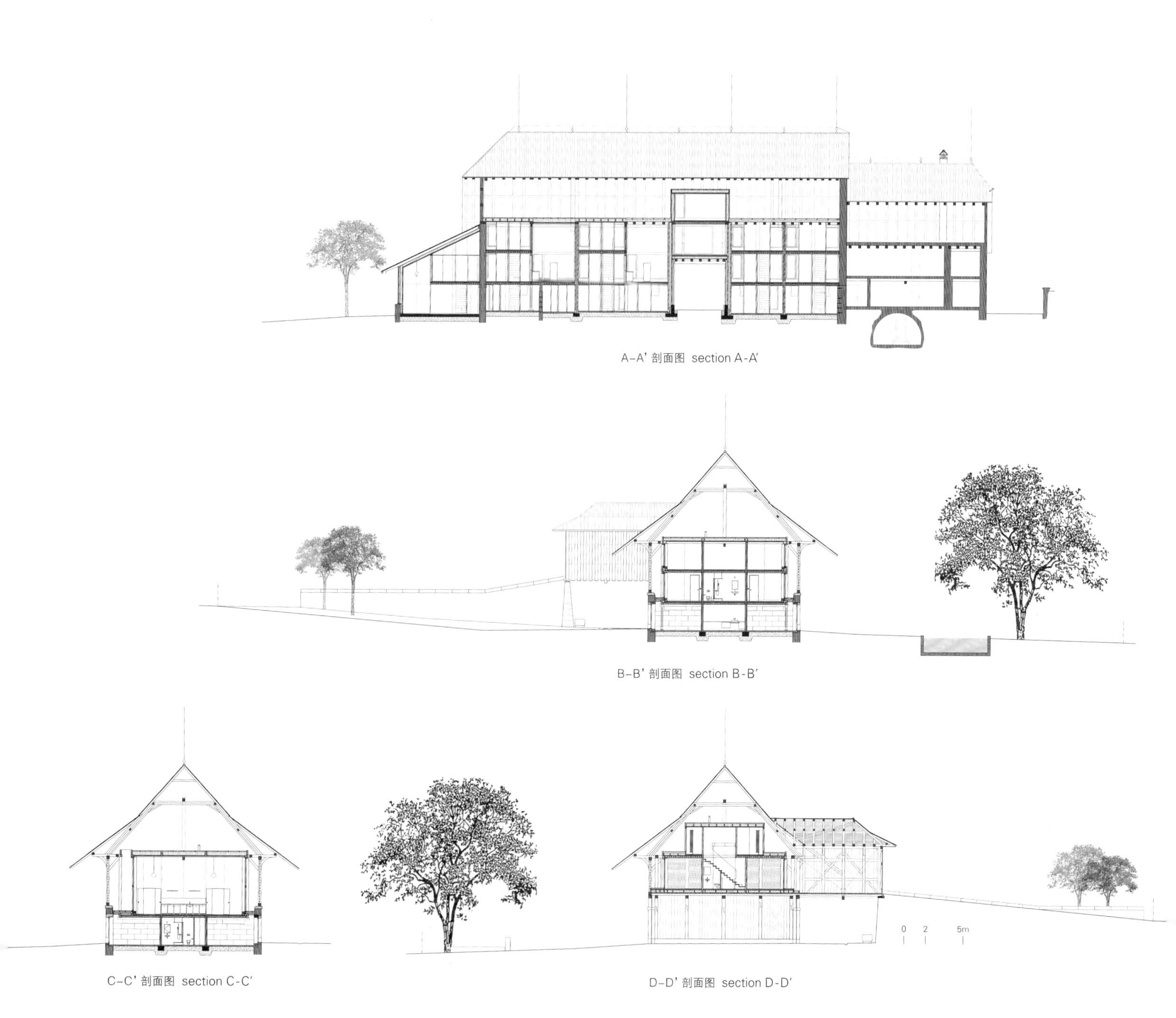

A–A' 剖面图 section A-A'

B–B' 剖面图 section B-B'

C–C' 剖面图 section C-C'

D–D' 剖面图 section D-D'

项目名称：Münchenwiler Castle – Barn and Residential Building / 地点：Kühergasse 4, Münchenwiler, Ber), Switzerland
事务所：bernath + widmer / 结构工程公司：Indermühle AG / 建筑设备工程公司：enerconom AG / 景观设计：Andreas Geser Landschaftsarchitekten AG / 木艺：Kühni AG, Ramsei and Loacker Holzbau / 用途：former use – agricultural use; now – residential use with ateliers
建筑面积：barn – 517m², residential building – 179m² / 总楼面面积：barn – 1,240m², residential building – 358m² /
竣工时间：2018 / 摄影师：©Roland Bernath (courtesy of the architect)

P16 Lundgaard & Tranberg Arkitekter

Was established in Copenhagen, 1985 by Boje Lundgaar and Lene Tranberg. Is founded on the development and realization of original architectural concepts, elevating ea individual project from the ordinary to innovative, visionary and poetic architecture. Tries to maximize architectur quality and value from a holistic viewpoint, with consideration for the inherent potential and limitations of each pr ect. Is shaped by the Nordic architectural tradition in whi humanism, craftsmanship and simplification are regarded as central virtues and where architecture engages in an enriching collaboration with culture, light and landscape.

P36 O-office Architects

s a Guangzhou-based practice established by HE Jianxiang[left] and JIANG Ying[right] in 2007. HE Jianxiang received his M. Arch(2009) and M. S. in Arch(2010) in KU Leuven, Belgium. JIANG Ying was granted the President Chirac Scholarship in 2000. Completed her M. Arch thesis and received Architect D.P.L.G in École d'Architecture de Versailles. Motivated to explore new architectural possibilities in contemporary southern Chinese urban context, it has been esting their methodology in architectural activities in different scales. Parallel to their professional practice, the team ersists in using architectural design as a critical instrument for research on the spatial and economic reality, and struggles to maintain the balance of the two. Conservation of the urban history and study on new urban collective housing in the fast-developing Pearl River Delta region has been O-office's main design focuses.

P8 Gihan Karunaratne

Is a British architect and a graduate of Royal College of Arts and Bartlett School of Architecture. Has taught and lectured in architecture, Interior and Urban Design in UK, Sri Lanka and China. Writes and researches extensively o art, architecture and urban design. Has exhibited globally many times at Art and Architecture Biennales. Was the director of the Architecture Element of the Colombo Art Biennale 2016. Is a recipient of The Bovis and Architect Journal Award for Architecture from Royal Academy Summer Exhibition 1999 and was made a Fellow of Royal Soc ety of Arts for Architecture, Design and Education in 201

P108 Balbek Bureau

s an architectural interior design practice based in Kyiv, Ukraine which was founded by the Ukrainian architect Slava Balbek. Now is a team of 55 architects and designers who design spaces for businesses and people who stand behind them. For 12 years of experience in creating bespoke commercial and residential spaces, it has been building heir community of liveable spaces. This process starts with a deep analysis of business demands, ergonomics and unctionality. Reimagines the traditional approach to creating commercial spaces, with design expertise across a wide range of industries. It is focused on innovation, uniqueness and effective performance to make one of a kind

P54 PSLA Architekten

Is a Vienna-based studio operating in the fields of contemporary architecture, urban planning, interior design and product design. Was co-founded by architects Lilli Pschill and Ali Seghatoleslami in 2013.
Lilli Pschill (1973, Vienna) attained her degree in architecture at the Technical University of Vienna. Ali Seghatoleslami (1975, Tehran) graduated from the Technical Universit in Vienna and studied at the Architetural Association in London, UK

P86 Eric Reeder

Is a licensed California architect whose career seeks to merge adaptation and urban speculation; whereby architecture is integral within larger contextual and social systems. For nearly a decade Eric practiced and taught in Seoul, South Korea. In 2018, Eric founded ERa in San Francisco, California. Currently lectures in the College of Environmental Design at UC Berkeley. His academic research and professional projects focus on housing, sustainable urbanism and adaptive reuse strategies. Eric studied urban design in Prague, Czech Republic through the University of Colorado prior to receiving a Bachelor of Environmental Design. He earned his M.Arch from UC Berkeley. In 2015 he published ***Adaptations Seoul*** and has received design awards and published recognition in North America, Europe and East Asia.

P158 Anna Roos

Studied architecture at the University of Cape Town and holds a postgraduate degree from the Bartlett School of Architecture at University College London. Moving to Bern, Switzerland in 2000, she worked as an architect, designing projects in South Africa, Australia, and Scotland. Has been working as a freelance architectural journalist since 2007 and writes for ***C3, Ensuite Kultur Magazin, Kolt, The Visitor Magazine and Swisspearl Architecture Magazine***. She also copyedits books for numerous publishing houses in Germany and Switzerland including: Gestalten, Park Books, Birkhäuser, and gta publishers. Anna seeks to convey her passion for architecture in her writing about the discipline. Her first book, ***Swiss Sensibility. The Culture of Architecture in Switzerland*** was published by Birkhäuser Verlag (2017). Currently works at Markus Schietsch Architects in Zurich.

P164 Bevk Perović Arhitekti

Was founded by Matija Bevk[right] and Vasa Perović[left] in 1997. Matija Bevk was born in Ljubljana, Slovenia in 1972 and graduated in architecture from the University of Ljubljana, 1999. Has been nominated for the Iakov Chernikhov Prize in 2012. Vasa Perović was born in Belgrade, Serbia in 1965 and graduated in architecture from University of Belgrade, 1992. Received a Master's Degree from Berlage Institute in 1994. Has been a Professor at Faculty of Architecture, University of Ljubljana since 2010. They work, alongside with the international team of 15 young architects, on a diverse range of projects, in different European countries They have been awarded Mies van der Rohe Emerging Architect Award, Berliner Kunstpreis, Plećnik Prize, Piranesi Award, etc.

P4 Philip D. Plowright

Is a licenced architect, design theorist, and academic researcher with degrees in studio art, architecture and cognitive linguistics. Recent publications include ***Making Architecture Through Being Human*** (Routledge, 2020), ***Sustainability and the City: Urban Poetics and Politics***(Lexington Books, 2017), ***Revealing Architectural Design: Methods, Frameworks & Tools***(Routledge, 2014), and chapters on theoretical issues of wilderness and sustainable design in ***Architecture and Sustainability: Critical Perspectives for Integrated Design***(ACCO, Belgium, 2015). His award-winning design work has been published in various academic proceedings, journals and architectural magazines such as ***Boundaries: International Architectural Magazine***(Italy), ***Archnet-IJAR***(USA), ***Bauwelt***(Germany), ***Arkinka: Revista de Arquitectura, Diseno y Construccion*** (Spain). Currently Editor-in-Chief of ***ENQ***, the ARCC Journal of Architectural Research.

P194 Arkitema Architects

Poul Schülein[picture-above], born in 1952, graduated from Royal Academy of Fine Arts and School of Architecture Copenhagen in 1978. Joined Arkitema in 1996 and has been a partner since 2001. Is responsible for historic restoration work, international competitions on transformations and major cultural projects. Has been appointed an external examiner for the Academy Council by the Royal Academy of Fine Arts, School of Architecture. Received the Europa Nostra Awards and the Order of the Dannebrog in 2010. Recently, has been appointed Royal Inspector of Listed Buildings by the Danish Building and Property Agency for Inspectorate 2, 2012-2016.

P220 bernath+widmer

Roland Bernath and Benjamin Widmer established their practice in Zurich, 2007. Roland studied architecture at the Zurich University Winterthur. After graduation, has been working as architect and architectural photographer and taught at the Lucerne School of Engineering and Architecture. Ben studied architecture at the Zurich University Winterthur, the Berlin University of the Arts, and ETH Zurich. Currently teaches at the Zurich University of Applied Sciences. Thoroughly investigating the characteristics of materials and exploring their structural use as well as attaching importance to quality craftsmanship have been a constant in the work of bernath+widmer.

bernath+widmer

P144 AMAA

Was created in 2012 by Marcello Galiotto[right] and Alessandra Rampazzo[left] based on their experience of working alongside Massimo Carmassi and Sou Fujimoto. They both graduated at Università Iuav di Venezia in 2010 and they completed their PhD in Architectural Design (Galiotto, 2015) and History of Architecture (Rampazzo, 2017). The restoration of the Santa Croce Convent in Venice was awarded with the honorable mention at the Cappochin Award 2015 (regional section) and the renovation of The [B] Zone was selected among the 10 best regional works in the Cappochin Award 2017. From 2015 AMAA has a branch office in Arzignano (Vi), in addiction to the main office in Venice.

P194 Christoffer Harlang

Was born in Copenhagen, 1958. Received a diploma in architecture at the Royal Academy of Fine Arts School of Architecture in 1983. Has a post graduate studies at Architectural Association School of Architecture London in 1985~86. Worked at British Council and Accademia di Danimarca in Rome as a fellow. Received a PhD on a thesis on Alvar Aalto and the legacy of Scandinavian. In 1998, he established his own office Christoffer Harlang Architects.

P180 EAA – Emre Arolat Architecture

Was founded in 2004 by Emre Arolat and Gonca Paşolar in Istanbul, as the continuation of Emre Arolat's architectural practices which he started at his parents' office, Arolat Architects, as an associate designer in 1987. Since its establishment, EAA has turned out to be one of the largest architectural offices in Turkey with a team of more than 50 people, with the reintegration of parents as partners. The firm continues its practice at the offices in Istanbul and London with projects that range from mixed-use buildings occupying millions of square meters, to a small place of worship counting only five hundred square meters.

P122 WMB studio

Ia a design studio based in Liverpool, England. Ed Butler studied Architecture at Liverpool John Moores University, receiving the RIBA President's Medals Komfort Award for his diploma thesis project. Prior to co-founding WMB studio, he worked for the Richard Rogers Partnership in London and BDP in Manchester. Ed has also taught BA Architecture for a number of years at his alma mater. Daniel Wiltshire studied Architecture at the Universities of Bath, Sheffield and KTH in Stockholm. Prior to co-founding WMB Studio he worked for Feilden Clegg Bradley Studios in Bath, and de Rijke Marsh Morgan (dRMM) and Allford Hall Monaghan Morris (AHMM) in London. Alongside WMB, Daniel teaches design studio at the University of Liverpool. Regardless of scale or budget, our projects are focused on creating a beautiful, functional and enriching experience for those who encounter them.

Chartier Dalix Architectes

P208 Chartier Dalix Architectes

Since its founding in 2008 by Frédéric Chartier[right] and Pascale Dalix[left], it has delivered some fifteen buildings. Is composed of 60 people and currently led by 3 partners of Frédéric Chartier, Pascale Dalix, and Sophie Deramond. Was awarded the Première Œuvre du Moniteur prize in 2009, and the '40 under 40' European prize for young architects in 2012. The residence hall at the Porte des Lilas was nominated for the Equerre d'argent prize and the Mies van der Rohe Award in 2014. The office recently received the 'Le Soufaché' prize, awarded by the French Academy of Architecture in recognition of its work as a whole.

图书在版编目(CIP)数据

都市校园 / 丹麦BIG建筑事务所编 ; 司炳月, 张晗, 王晓华译. — 大连 : 大连理工大学出版社, 2021.3
ISBN 978-7-5685-2936-5

Ⅰ. ①都… Ⅱ. ①丹… ②司… ③张… ④王… Ⅲ. ①学校－教育建筑－建筑设计－案例 Ⅳ. ①TU244

中国版本图书馆CIP数据核字(2021)第019243号

出版发行：大连理工大学出版社
（地址：大连市软件园路80号　　邮编：116023）
印　　刷：上海锦良印刷厂有限公司
幅面尺寸：225mm×300mm
印　　张：14.75
出版时间：2021年3月第1版
印刷时间：2021年3月第1次印刷
出 版 人：苏克治
统　　筹：房　磊
责任编辑：杨　丹
封面设计：王志峰
责任校对：张昕焱
书　　号：978-7-5685-2936-5
定　　价：298.00元

发　行：0411-84708842
传　真：0411-84701466
E-mail：12282980@qq.com
URL：http://dutp.dlut.edu.cn

本书如有印装质量问题，请与我社发行部联系更换。

墙体设计
ISBN：978-7-5611-6353-5
定价：150.00 元

新公共空间与私人住宅
ISBN：978-7-5611-6354-2
定价：150.00 元

住宅设计
ISBN：978-7-5611-6352-8
定价：150.00 元

老年住宅
ISBN：978-7-5611-6569-0
定价：150.00 元

小型建筑
ISBN：978-7-5611-6579-9
定价：150.00 元

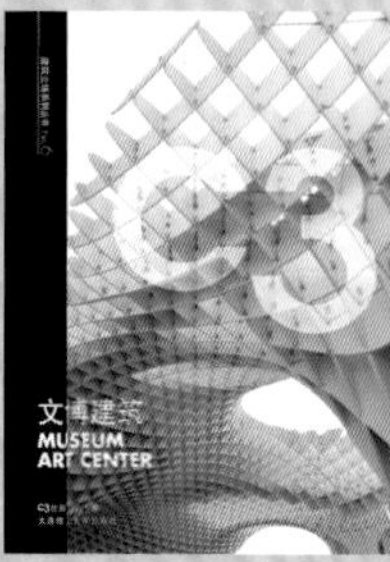

文博建筑
ISBN：978-7-5611-6568-3
定价：150.00 元

流动的世界：日本住宅空间设计
ISBN：978-7-5611-6621-5
定价：200.00 元

创意运动设施
ISBN：978-7-5611-6636-9
定价：180.00 元

墙体与外立面
ISBN：978-7-5611-6641-3
定价：180.00 元

空间与场所之间
ISBN：978-7-5611-6650-5
定价：180.00 元

文化与公共建筑
ISBN：978-7-5611-6746-5
定价：160.00 元

城市扩建的四种手法
ISBN：978-7-5611-6776-2
定价：180.00 元

复杂性与装饰风格的回归
ISBN：978-7-5611-6828-8
定价：180.00 元

企业形象的建筑表达
ISBN：978-7-5611-6829-5
定价：180.00 元

图书馆的变迁
ISBN：978-7-5611-6905-6
定价：180.00 元

亲地建筑
ISBN：978-7-5611-6924-7
定价：180.00 元

旧厂房的空间蜕变
ISBN：978-7-5611-7093-9
定价：180.00 元

混凝土语言
ISBN：978-7-5611-7136-3
定价：228.00 元

建筑入景
ISBN：978-7-5611-7306-0
定价：228.00 元

新医疗建筑
ISBN：978-7-5611-7328-2
定价：228.00 元

内在丰富性建筑
ISBN：978-7-5611-7444-9
定价：228.00 元

建筑谱系传承
ISBN：978-7-5611-7461-6
定价：228.00 元

伴绿而生的建筑
ISBN：978-7-5611-7548-4
定价：228.00 元

大地的皱折
ISBN：978-7-5611-7649-8
定价：228.00 元

在城市中转换
ISBN：978-7-5611-7737-2
定价：228.00 元

锚固与飞翔——挑出的住居
ISBN：978-7-5611-7759-4
定价：228.00 元

创造性加建：我的学校，我的城市
ISBN：978-7-5611-7848-5
定价：228.00 元

文化设施：设计三法
ISBN：978-7-5611-7893-5
定价：228.00 元

终结的建筑
ISBN：978-7-5611-8032-7
定价：228.00 元

博物馆的变迁
ISBN：978-7-5611-8226-0
定价：228.00 元

微工作·微空间
ISBN：978-7-5611-8255-0
定价：228.00 元

居住的流变
ISBN：978-7-5611-8328-1
定价：228.00 元

本土现代化
ISBN：978-7-5611-8380-9
定价：228.00 元

气候与环境
ISBN：978-7-5611-8501-8
定价：228.00 元

能源与绿色
ISBN：978-7-5611-8911-5
定价：228.00 元